T0207286

Communications in Computer and Information Science　　2129

Rationale

The CCIS series is devoted to the publication of proceedings of computer science conferences. Its aim is to efficiently disseminate original research results in informatics in printed and electronic form. While the focus is on publication of peer-reviewed full papers presenting mature work, inclusion of reviewed short papers reporting on work in progress is welcome, too. Besides globally relevant meetings with internationally representative program committees guaranteeing a strict peer-reviewing and paper selection process, conferences run by societies or of high regional or national relevance are also considered for publication.

Topics

The topical scope of CCIS spans the entire spectrum of informatics ranging from foundational topics in the theory of computing to information and communications science and technology and a broad variety of interdisciplinary application fields.

Information for Volume Editors and Authors

Publication in CCIS is free of charge. No royalties are paid, however, we offer registered conference participants temporary free access to the online version of the conference proceedings on SpringerLink (http://link.springer.com) by means of an http referrer from the conference website and/or a number of complimentary printed copies, as specified in the official acceptance email of the event.

CCIS proceedings can be published in time for distribution at conferences or as post-proceedings, and delivered in the form of printed books and/or electronically as USBs and/or e-content licenses for accessing proceedings at SpringerLink. Furthermore, CCIS proceedings are included in the CCIS electronic book series hosted in the SpringerLink digital library at http://link.springer.com/bookseries/7899. Conferences publishing in CCIS are allowed to use Online Conference Service (OCS) for managing the whole proceedings lifecycle (from submission and reviewing to preparing for publication) free of charge.

Publication process

The language of publication is exclusively English. Authors publishing in CCIS have to sign the Springer CCIS copyright transfer form, however, they are free to use their material published in CCIS for substantially changed, more elaborate subsequent publications elsewhere. For the preparation of the camera-ready papers/files, authors have to strictly adhere to the Springer CCIS Authors' Instructions and are strongly encouraged to use the CCIS LaTeX style files or templates.

Abstracting/Indexing

CCIS is abstracted/indexed in DBLP, Google Scholar, EI-Compendex, Mathematical Reviews, SCImago, Scopus. CCIS volumes are also submitted for the inclusion in ISI Proceedings.

How to start

To start the evaluation of your proposal for inclusion in the CCIS series, please send an e-mail to ccis@springer.com.

Vladimir M. Vishnevskiy ·
Konstantin E. Samouylov · Dmitry V. Kozyrev
Editors

Distributed Computer and Communication Networks

26th International Conference, DCCN 2023
Moscow, Russia, September 25–29, 2023
Revised Selected Papers

 Springer

Editors
Vladimir M. Vishnevskiy ⓘD
V.A.Trapeznikov Institute of Control
Sciences of Russian Academy of Sciences
Moscow, Russia

Konstantin E. Samouylov ⓘD
Peoples' Friendship University of Russia
Moscow, Russia

Dmitry V. Kozyrev ⓘD
Department of Sciences
Peoples' Friendship University of Russia
Moscow, Russia

V.A.Trapeznikov Institute of Control
Sciences of Russian Academy of Sciences
Moscow, Russia

ISSN 1865-0929   ISSN 1865-0937 (electronic)
Communications in Computer and Information Science
ISBN 978-3-031-61834-5   ISBN 978-3-031-61835-2 (eBook)
https://doi.org/10.1007/978-3-031-61835-2

This Springer imprint is published by the registered company Springer Nature Switzerland AG
The registered company address is: Gewerbestrasse 11, 6330 Cham, Switzerland

If disposing of this product, please recycle the paper.

Preface

This volume contains a collection of revised selected full-text papers presented at the 26th International Conference on Distributed Computer and Communication Networks (DCCN 2023), held in Moscow, Russia, during September 25–29, 2023. DCCN 2023 was jointly organized by the Russian Academy of Sciences (RAS), the V.A. Trapeznikov Institute of Control Sciences of RAS (ICS RAS), the Peoples' Friendship University of Russia (RUDN University), the National Research Tomsk State University, and the Institute of Information and Communication Technologies of the Bulgarian Academy of Sciences (IICT BAS).

The conference was a continuation of the traditional international conferences of the DCCN series, which have taken place in Sofia, Bulgaria (1995, 2005, 2006, 2008, 2009, 2014); Tel Aviv, Israel (1996, 1997, 1999, 2001); and Moscow, Russia (1998, 2000, 2003, 2007, 2010, 2011, 2013, 2015–2022) in the last 26 years. The main idea of the conference was to provide a platform and forum for researchers and developers from academia and industry from various countries working in the area of theory and applications of distributed computer and communication networks, mathematical modeling, and methods of control and optimization of distributed systems, by offering them a unique opportunity to share their views, as well as discuss prospective developments and pursue collaboration in this area. The content of this volume is related to the following subjects:

- 5G/6G communication networks algorithms and protocols
- Analytical modeling and simulation of communication systems
- Computer and telecommunication networks control and management
- Wireless and mobile networks
- High-altitude telecommunications platforms
- Performance analysis, QoS/QoE evaluation, and network efficiency
- Queuing theory and reliability theory applications
- Cloud computing, distributed and parallel systems
- Centimeter- and millimeter-wave radio technologies

The DCCN 2023 conference received 122 submissions from authors from 18 different countries. From these, 105 high-quality papers in English were accepted and presented during the conference. All submissions underwent a rigorous double-blind peer-review process with 3 reviews per submission. The current volume contains 8 extended papers which were recommended by session chairs and selected by the Program Committee for the Springer CCIS post-proceedings.

All the papers selected for the post-proceedings volume are given in the form presented by the authors. These papers are of interest to everyone working in the field of computer and communication networks.

We thank all the authors for their interest in DCCN, the members of the Program Committee for their contributions, and the reviewers for their peer-reviewing efforts.

September 2023

Vladimir M. Vishnevskiy
Konstantin E. Samouylov
Dmitry V. Kozyrev

Organization

Program Committee Chairs

V. M. Vishnevskiy (Chair)   ICS RAS, Russia
K. E. Samouylov (Co-chair)   RUDN University, Russia

Publication and Publicity Chair

D. V. Kozyrev   ICS RAS and RUDN University, Russia

International Program Committee

S. M. Abramov	Program Systems Institute of RAS, Russia
A. M. Andronov	Transport and Telecommunication Institute, Latvia
T. Atanasova	IICT BAS, Bulgaria
S. E. Bankov	Kotelnikov Institute of Radio Engineering and Electronics of RAS, Russia
A. S. Bugaev	Moscow Institute of Physics and Technology, Russia
S. R. Chakravarthy	Kettering University, USA
D. Deng	National Changhua University of Education, Taiwan
S. Dharmaraja	Indian Institute of Technology, Delhi, India
A. N. Dudin	Belarusian State University, Belarus
A. V. Dvorkovich	Moscow Institute of Physics and Technology, Russia
D. V. Efrosinin	Johannes Kepler University Linz, Austria
Yu. V. Gaidamaka	RUDN University, Russia
Yu. V. Gulyaev	Kotelnikov Institute of Radio-engineering and Electronics of RAS, Russia
V. C. Joshua	CMS College Kottayam, India
H. Karatza	Aristotle University of Thessaloniki, Greece
N. Kolev	University of São Paulo, Brazil
G. Kotsis	Johannes Kepler University Linz, Austria

Organizing Committee

D. V. Kozyrev (Publication and   ICS RAS and RUDN University, Russia
   Publicity Chair)
A. A. Larionov   ICS RAS, Russia
Y. S. Aleksandrova   ICS RAS, Russia
S. P. Moiseeva   Tomsk State University, Russia
T. Atanasova   IICT BAS, Bulgaria
I. A. Kochetkova   RUDN University, Russia

Organizers and Partners

Organizers

Russian Academy of Sciences (RAS), Russia
V.A. Trapeznikov Institute of Control Sciences of RAS, Russia
RUDN University, Russia
National Research Tomsk State University, Russia
Institute of Information and Communication Technologies of the Bulgarian Academy
of Sciences, Bulgaria
Research and Development Company "Information and Networking Technologies",
Russia

Support

Information support was provided by the Russian Academy of Sciences. The conference
was organized with the support of the IEEE Russia Section, Communications Society
Chapter (COM19) and the RUDN University Strategic Academic Leadership Program.

Contents

Modeling and Comparison of Different Management Approaches on the Intersections Network

Timofei I. Tislenko[1], Daria V. Semenova[1]([✉]), Aleksandr A. Soldatenko[1],
Nataly A. Sergeeva[2], Elena E. Goldenok[3], and Nadezhda V. Kononova[4]

[1] School of Mathematics and Computer Science, Siberian Federal University,
Krasnoyarsk, Russian Federation
{DVSemenova,ASoldatenko}@sfu-kras.ru
[2] Analytics GroupsLLC "R.D.Science", Krasnoyarsk, Russian Federation
info@rd-science.com
[3] Department of Medical Cybernetics and Informatics Prof. V.F. Voino-Yasenetsky,
Krasnoyarsk, Russian Federation
[4] School of Space and Information Technology, Siberian Federal University,
Krasnoyarsk, Russian Federation
NKoplyarova@sfu-kras.ru

Abstract. The aim of the work is to develop a set of algorithms and programs for simulation modeling of traffic flows and traffic optimization for sections of the Krasnoyarsk road network. The paper investigates models of the process of selecting the phases of traffic lights in order to minimize the travel time of vehicles through the road network section. As a way to control traffic light objects, a phase selection process has been proposed based on the Multiagent Reinforcement Learning for Integrated Network also known as MARLIN. As an alternative to the traffic light phases change process in MARLIN, a fuzzy model of phase duration change on the next cycle is proposed. A microscopic Intelligent Driver Model (IDM) was used to simulate road traffic. To compare the models, a series of computational experiments were performed on the example of a section of the Krasnoyarsk road network.

Keywords: multi-agent reinforcement learning · Q-learning · Markov decision process · fuzzy-logic · traffic light control system

1 Introduction

Currently, for large cities, one of the most pressing problems is reducing the capacity of road networks as a consequence of their strong branching and increased traffic due to growth number of vehicles. The old traffic light control

This work is supported by the Krasnoyarsk Mathematical Center and financed by the Ministry of Science and Higher Education of the Russian Federation (Agreement No. 075-02-2023-936).

strategies used most often, such as controlled schedule management, became ineffective of the lack or slow adaptation to the traffic flow variability during the day (the pendulum flow behavior). One solution to this problem is the choice of traffic control optimal modes. In this regard, the use of simulation tools to solve traffic problems allows to evaluate the current traffic conditions and propose specific solutions to optimize the urban traffic. Three main classes of mathematical models can be identified for the transport networks and flows analysis [1]: predictive models for estimating the volume of transport between different objects producing transport flows; models of the transport flows dynamics and optimization models. In this paper, we consider models of two of the above classes: a microscopic type model for traffic simulation; a multi-agent learning model with reinforcement; and a fuzzy logic controller model for the traffic lights adaptive control.

The software and mathematical models for the transport networks analysis, are sufficiently developed and widely used to solve the transport problems of large urban [2–7]. In particular, the following software products for the simulation of traffic and regulation of urban traffic flows available to us were considered GPSS, Anylogic PLE, MATSim, PTV VISSIM (Table 1).

Table 1. Comparison of the selected traffic simulation systems.

Solution	Code debug	Main limitations	Simulation approach	Programming language
GPSS	yes	Unlimited	external	GPSS World
Anylogic	no	Model time less than 1 h, no compatibility with 14ver	micro	java
MATSim	yes	Unlimited	micro	java
PTV VISSIM	no	Model grid less than 1×1 km, model time less than 10 min	any	no

In our view, the commercial packages Anylogic [3], PTV VISSIM [4] for solving modeling problems are unsuitable for some significant reasons: projects from the old version are incompatible, the model time is limited, fine-tuning and debugging the program code is absent, the ability to edit the source code in a full editor is absent too. The standard traffic simulation module (mobsim) for the software package MATSim adopts the computationally efficient queue-based approach [5] and there it is distinguished between

- physical simulations, featuring detailed car following,
- cellular automata, in which roads are discretized into cells,
- queue-based simulations, where traffic dynamics are modeled with waiting queues,
- mesoscopic models, using aggregates to determine speeds,
- macroscopic models, based on flows rather than single traveler units (e.g., cars).

Therefore, one can use the listed traffic models only if the "mobsim" module is modified. Using the system GPS [6] for construction and implementation of a traffic simulation model is comparable in labor costs with writing the own original package of software. Another significant disadvantage of using off-the-shelf solutions is their isolation, which makes it difficult to create a custom environment for the OpenAI Gym [7]. Thus, the most characteristic problems are: the inability to integrate the code with third-party information systems and traffic flow sensors; the lack of open API and algorithms for processing this data to compute optimization procedures for the next phase of traffic lights and network of traffic light objects. The analysis of available free software revealed the feasibility of developing and applying the own set of tools for simulation modeling of traffic flows and optimization of road traffic for some road network sections in Krasnoyarsk.

2   Task Description

The mathematical description of the used models of the traffic flow dynamics and of adaptive control of the traffic light objects are given in this section. Mathematical models of traffic flows are commonly divided into three large categories micro, meso and macro-level [1,2] according to the level of detail modeling objects. To simulate traffic, we constructed the microscopic type model (Intellectual Driver Model, ID) [8]. The model is presented here. Its description is given in Subsect. 2.1.

A phase selection process based on multi-agent learning with reinforcement has been proposed as a way of controlling of the traffic light objects controlling (Multi-agent Reinforcement Learning for Integrated Network, MARLIN) [9–11]. The MARLIN problems for its effective application require solving the problems of aggregate agent control [12,13] and increasing the dimensionality of matrices [14] when extending of the road network coverage. It should be noted that mutual interlocks and conflicts of control of the agents separately are resolved at the aggregate management. A description of the MARLIN is provided in Subsect. 2.2.

As an alternative to the phase change process, the changing the phase duration on the next cycle can be considered. The operation of the traffic light in normal mode is characterized by a constant duration of the green and red light and the entire cycle. To increase the control flexibility of the network of traffic lights it is necessary to change the duration of the elements of the traffic light cycle in accordance with the number of vehicles coming to the intersection.

For these purposes, it is proposed to organize the operation of traffic light objects on the fuzzy logic principles [15–17]. In the proposed fuzzy model, the duration of traffic light modes should vary depending on time of day and traffic intensity, and the length of the traffic light cycle, determined by the total duration of modes, should not exceed 300 s. The fuzzy control model of the traffic light object is presented in Subsect. 2.3.

2.1 Intellectual Driver Model

The Intelligent Driver Model (IDM) developed by Treiber at al. [8] is recognized as one of the most successful microscopic models. At the micro level, the equations of motion for each vehicle can be written out explicitly. Vehicles are viewed as individual entities with their own characteristics and behavior. The acceleration of each car is described as a function of its speed, of the distance to the car ahead and of the speed relative to the leader (Fig. 1).

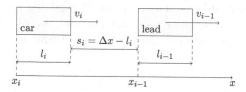

Fig. 1. An example of a microscopic model.

The equation of motion of the i-th car in classical IDM notations is [8]

$$\frac{dv_i}{dt} = a_{\text{free}} + a_{\text{deceleration}} = a_i \left(1 - \left(\frac{v_i}{v_{0,i}} \right)^\delta \right) - a_i \left(\frac{s^* (v_i \Delta v_i)}{s_i} \right)^2,$$

$$s^* (v_i \Delta v_i) = s_{0,i} + v_i T_i + \frac{v_i \Delta v_i}{\sqrt{2 a_i b_i}},$$

(1)

where

- $s_{0,i}$ is the minimum desired distance between i and $i - 1$,
- $v_{0,i}$ is the maximum desired speed i,
- δ is the acceleration component responsible for "smoothness",
- T_i is the reaction time of the i-th driver,
- a_i is the maximum acceleration i,
- b_i is the comfortable braking i,
- s^* is the possible distance between i and $i - 1$.

The model of the road network is given by the multigraph. The unique number is assigned to each vertice. The edges store information about the connected vertices, its coordinates, the length of the road, and the number of lanes. The road transitioning from the intersection is added to the tail of the queue. The road transitioning and the traffic lights are described by the introduction of special zones (Fig. 2):

- Slow down zone: characterized by a slow down distance and a slow down factor, is a zone in which vehicles slow down their maximum speed using the slow down factor $v_{0,i} = \theta v_{0,i}, 0 < \theta < 1$;

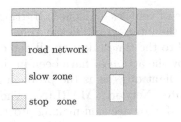

Fig. 2. Example of zone placement at an intersection. The Slow down zone is high-lighted in yellow, and the Stop zone is highlighted in red. (Color figure online)

– Stop zone: characterized by a stop distance, is a zone in which vehicles stop. This is achieved using a damping force through this dynamics equation
$$v_{0,i} = -b_i \frac{v_i}{v_{0,i}}.$$

The calibration procedure and the numerical experiments with this model showed that its properties are stable to parameter variations; the model demonstrates the realistic behavior at the acceleration and the deceleration and reproduces the main observed traffic flow properties.

2.2 MARLIN for Traffic Light Network

Let vehicles drive into each controlled intersection. The traffic light object located at the intersection is considered as an agent. The environment is considered a section of the road network containing controlled intersections and adjacent road sections starting 70 m before the stop lines of the intersection. On each section of the road there are detectors that record the entry and the exit time. The agent state is the number of the active phase of the traffic light object. The traffic lights objects phases change in sequence.

Let sets of environment states $S^0, \ldots, S^K$ and sets of agents decisions $A^0, \ldots, A^K$ be described for each of K traffic light objects. The $\mathbf{s}_t \in S^0 \times \cdots \times S^K = \mathbf{S}$ is denoted the state of the environment at the moment time t, and the $\mathbf{a}_t \in A^0 \times \cdots \times A^K = \mathbf{A}$ is denoted the control of the traffic lights phases changes at the time t. We assume that for the k-th traffic light, the set S^k is the residue class ring of integers, $\mathbb{Z}_{n_k} = \langle \mathbb{Z}, +, \cdot \rangle$, where $|S^k| = n_k$ is the number of classes $\mathbb{Z}_{n_k}$. The characteristic of the residue ring $\mathbb{Z}_{n_k}$ is n_k and in this model reflects the number of phase changes of the k-th traffic light before it returns to its initial value, hence the number of actions $|\mathbf{a}_k|$ before returning to the initial state of the k-th traffic light is n_k. Consequently, if the environment is in the state $\mathbf{s}$, then by choosing the action $\mathbf{a} = \delta(\mathbf{s})$ the new state $\mathbf{s}'$ is defined by the formula $\mathbf{s}' = (\mathbf{a} + \mathbf{s}) \mod |\mathbf{S}|$. Let assume that at the moment time t the environment, being in the state $\mathbf{s}_t \in \mathbf{S}$, expects the agents to make the actions $\mathbf{a}_t \in \mathbf{A}$, after which it takes a step, moving to the state $\mathbf{s}_{t+1}$ with the probability $p(\mathbf{s}_{t+1}|\mathbf{s}_t, \mathbf{a}_t)$. The conditional probabilities $p(\mathbf{s}_{t+1}|\mathbf{s}_t, \mathbf{a}_t)$ for all states $\mathbf{s}_t \in \mathbf{S}$ and actions $\mathbf{a}_t \in \mathbf{A}$ form the transition matrix of the Markov chain $\mathcal{P}$ [18]. For each road section, a queue of vehicles is defined, starting from the detector and ending at the stop line of the intersection. Let at time t the value of the reward function $r(\mathbf{s}_t, \mathbf{a}_t)$ be equal to the time taken by all detected cars to traverse the road segments activated by the action $\mathbf{a}_t$ have been in the state $\mathbf{s}_t$. We treat the K phase control of traffic light objects as the problem Multi-Agent Reinforcement Learning for Integrated Network (MARLIN). The problem is formulated as follows. Let be given: a Markov decision-making process for controlling a road network of the K traffic light objects, phases of the traffic light objects are active at the initial time moment. It is required to find traffic light control delivers the maximum value of the function [19]

$$V^* \left(\{\mathbf{s}_{t'}, \delta\}_{t'=0}^{t'=t} \right) = \max_{a \in \mathbf{A}} Q_t(\mathbf{s}_t, \mathbf{a}), \tag{2}$$

$$Q_t(\mathbf{s}_t, \mathbf{a}) = \sum_{s' \in S} p(\mathbf{s}'|\mathbf{s}_t, \mathbf{a}) \left(r(\mathbf{s}'|\mathbf{s}_t, \mathbf{a}) + \gamma \max_{\mathbf{a}' \in \mathbf{A}} Q_{t-1}(\mathbf{s}', \mathbf{a}') \right). \tag{3}$$

Here, the variable $0 < \gamma < 1$, is called the revaluation coefficient and shows how many times the deferred reward decreases in one step of time.

The problem of aggregate control of multiple traffic lights reduces to a single agent control problem with a learning function [9–11,20]

$$Q_t^k(\mathbf{s}_t, a_t^k) = \sum_{a^j \in A^j} p(\mathbf{s}_t, a^{kj}) Q_t(\mathbf{s}_t, a^{kj}), \tag{4}$$

where $a^{kj} = \left(a_t^k; a^j \right)$ is the coordinated action of agents k and j. In this case, we will search for the optimal control for a fixed traffic light k in form a solution to the Multi-Agent Reinforcement Learning problem with a condition coordination of the MARLIN agents in the form [20]

$$\mathbf{a}_t = \arg \max_{a^k \in A^k} \sum_{a^j \in A^j} Q_t^k(\mathbf{s}_t, a^k) p(\mathbf{s}'|\mathbf{s}_t, a^{kj}), \tag{5}$$

where $\mathbf{s}' = (\mathbf{a} + \mathbf{s}_t) \mod |S^0 \times \cdots \times S^K|$ and $p(\mathbf{s}'|\mathbf{s}_t, a^{kj})$ is the probability that agent j will choose action a^j given the current joint state st and agent k chosen action a^k.

This model was previously realized for a section of the road network of Krasnoyarsk consisting of two intersections [20]. Computational experiments series have been conducted in the Anylogic simulation modeling environment were described in [21,22]. In the [23], the problem of input parameters selection was considered. Section 3 will present the results of a new complex of simulation modeling programs in Python3 for the same section of the road network.

2.3   Fuzzy Logic Controller for Traffic Light

As an alternative to the traffic light phase switching process in MARLIN, we can consider changing the phase duration on the next cycle, which is successfully implemented by a fuzzy logic controller. A MISO (Multiple Input Single Output) type system with three inputs is developed for the traffic light control based on the Mamdani method [15–17]. The following linguistic variables are considered as inputs

– time of day with the term-set $\mathcal{DT} = \{$ "morning", "day", "evening", "night"$\}$;
– motion density with the term-set $\mathcal{V} = \{$ "run", "wait', "jam"$\}$;
– reward (standing time of cars at a traffic light in seconds) with the term-set $\mathcal{R} = \{$ "small", "medium", "large"$\}$.

The output linguistic variable of the controller is phase duration defined by the term-set $P=\{$ "long", "medium", "short"$\}$. The membership functions of the terms of the linguistic variables *day_time, vehicles, reward, phase* are shown in Fig. 3.

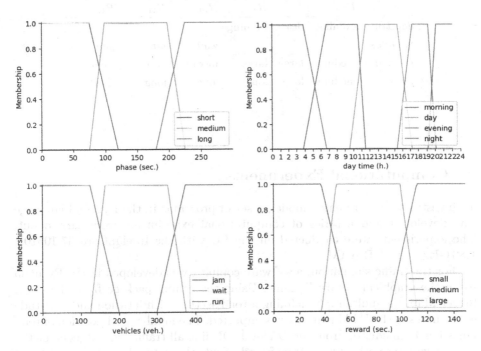

Fig. 3. Membership functions of the linguistic variables *day_time, vehicles, reward, phase*.

The Mamdani model is a set of rules, where each rule defines some fuzzy point in the specified space [24]. In this paper, the knowledge base is represented by twelve rules (Table 2) of the form:

R : IF $(day\_time = A)$ AND $(vehicles = B)$ AND $(reward = C)$
       THEN $(phase = D)$,

where $A \in \mathcal{DT}$, $B \in \mathcal{V}$, $C \in \mathcal{R}$, $D \in \mathcal{P}$.

Based on the set of fuzzy points, a fuzzy graph is formed, the interpolation mechanism between points depends on the used fuzzy logic apparatus. We use Zadeh's t-norm to formalize a fuzzy conjunction and the Mamdani implication [24] to implement a fuzzy implication.

Table 2. Knowledge base

	R_1	R_2	R_3	R_4	R_5	R_6
day_time	morning	night	–	–	day	day
vehicles	–	–	–	run	jam	–
reward	–	–	small	–	medium	medium
phase	short	short	short	short	short	medium
	R_7	R_8	R_9	R_{10}	R_{11}	R_{12}
day_time	evening	day	evening	–	–	–
vehicles	–	–	–	wait	jam	jam
reward	medium	large	large	medium	large	small
phase	medium	long	long	long	long	short

3   Computational Experiments

To investigate the presented model, a set of programs in the Python3 language was developed and a series of computational experiments were carried out. The experiments were conducted on the PC with the Intel®Core™i7-10510U 1.80 GHz and 8 GB RAM.

For the traffic simulation, a software complex was developed in the Python3 using the graphical library Pygame. Using the osmnx package from the Open Street Map, the multigraph, defining a road section with latitude and longitude coordinates in the UTM markup was imported. The number and position of road lanes has been edited. From the DATA.JSON file, all traffic light objects for the respective intersections were transferred. Next, the network was imported and transport routes were built on it using the network and threading packages.

For modeling and subsequent training of the traffic light objects a section of the Krasnoyarsk road network with two intersections Svobodny Ave and Lesoparkovaya St. was chosen (Fig. 4).

The control algorithm of the traffic light object is shown in Fig. 5. In simulation modeling, as a result of calling the generator procedure, vehicles are created in quantities close to real values. The vehicles then travel on the road network until they leave its coverage area. When a machine enters a detected road network segment, the VEHICLE DETECTOR module adds pairs consisting of pointers

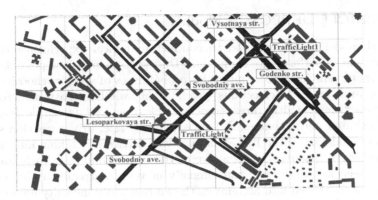

Fig. 4. The model of the section of the Krasnoyarsk road network consisting of two intersections Svobodny Ave and Lesoparkovaya St.

to the vehicle object and the current model time to one of the tcf (time collection forward) collections. Further, cars are removed from the collection when passing through an intersection.

In the DESCISION MODULE, the period variable shows the time during which the QLEARNING event is called to calculate the control according to (5) or FLCONTROL depending on the model selection. Based on DESCISION MODULE outputs, a decisions to switch the phases of traffic lights the TrafficLight for the intersection Svobodny Ave. – Vysotnaya St. and the TrafficLight1 for the intersection Svobodny Ave. – Lesoparkovaya St. (Fig. 5) are made.

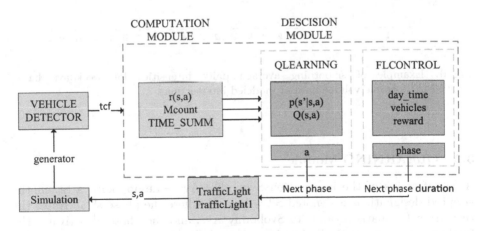

Fig. 5. Algorithm of control of traffic light objects.

3.1   VEHICLE DETECTOR Module

The IDM model is implemented to conduct numerical experiments comparing management performance indicators. During modeling, it is possible to set parameters of vehicle sources (intensity, probabilities of vehicles choosing given directions and lanes), intersections (traffic light switching patterns), roadway (braking coefficient, width, presence of obstacles), as well as vehicle characteristics (maximum and recommended speed, vehicle size, driver reaction speed). Simulation modeling of the flow dynamics was carried out taking into account the following conditions: vehicles arrive at the intersection from each lane from the five directions considered; the intensity of vehicle arrivals at the modeled intersections was taken into account as the average daily intensity for weekdays based on the data on traffic intensity at these intersections for 2017–2019 (Fig. 6). For model (1) following vehicle parameters are given:

$$s_0 = 4\,m, \quad v_0 = 16.6\,\frac{m}{s}, \quad T = 1\,s, \quad a = 1.44\,\frac{m}{s^2}, \quad b = 4.6\,\frac{m}{s^2}.$$

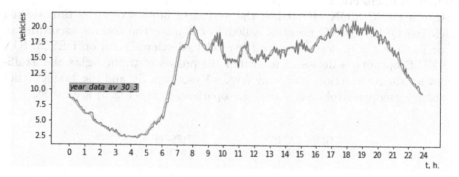

Fig. 6. Example of smoothing average daily intensities for weekdays based on 2017–2019 traffic volume data at modeled intersections.

3.2   QLEARNING Module

A set of states of the intersection Svobodny Ave.—Lesoparkovaya St. in the adopted designations is denoted $S^0 = \{s^0, s^1\}$ where the phase s^0 activates the turn from Lesoparkovaya St. to Svobodny Ave. and the phase s^1 activates the further movement on Svobodny Ave. $A^1 = \{a^0, a^1\}$ is a set of actions for this intersection. For the intersection Svobodniy Ave—Godenko St. the set of states is $S^1 = \{s^0, s^1, s^2\}$, where phase s^0 activates the movement on Godenko St., s^1 activates the movement on Svobodniy Ave., s^2 activates the movement on Vysotnaya St.. Set of actions for the intersection of Svobodniy Ave.—Godenko St. is $A^1 = \{a^0, a^1, a^2\}$, where a^j is interpreted as "change phase j times".

For the MARL model, in the absence of agent coordination, the control is obtained as a solution to the reinforcement learning problem for each traffic light separately by the (5).

The MARLIN control is obtained as a solution to a reinforcement learning problem with the condition of coordinating traffic lights in the network. The transition graph between the numbered states is shown in Fig. 7.

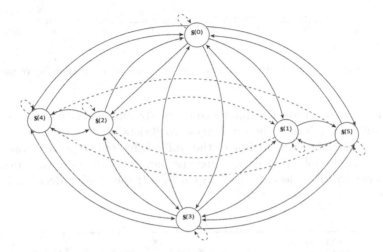

Fig. 7. Transition graph for the network of two multiphase traffic light objects of the road network section consisting of the following intersections: Svobodny Ave.— Lesoparkovaya St., Svobodny Ave.—Vysotnaya St.

Adjustable parameters of MARL and MARLIN models: γ is revaluation coefficient and $p(\mathbf{s}'|\mathbf{s}_t, \mathbf{a})$ is conditional probabilities in (3). The parameter fitting mechanism is explained as follows [23].

The sequence of values accepted by the (3) under the condition can be rewritten as

$$Q_{t+1} = (1 - \mathbf{a}_t)^t Q_1 + \sum_{i=1}^{t} \mathbf{a}_t (1 - \alpha_t)^{t-i} R_i, \tag{6}$$

where R_i is rewards received at time i, α_t are estimates of conditional probabilities $p(\mathbf{s}'|\mathbf{s}_t, a)$. It is not difficult to show [25] that

- if $0 < \alpha_t < 1$, then $Q_1 < Q_t < 0.25 \sum_{i=1}^{t} R_i$;
- if $\alpha_t = 0$, then $Q_{t+1} = Q_1$;
- if $\alpha_t - 1$, then $Q_{t+1} = R_t$.

Experiments have shown that the starting value of the conditional probability can be given by any number from $[10, 11]$. In particular, the resulting estimate α' from a past simulation session can be used. Figure 8 demonstrates a comparison of two graphs of the function $\{Q_t\}$ at $\alpha_0 = 0$ (blue) and at $\alpha_0 = \alpha'$ (red), and their absolute difference (green).

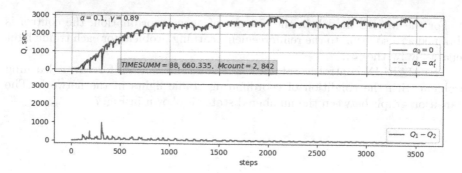

Fig. 8. Comparison of Q_t function at different values of α_0. (Color figure online)

Figure 9 demonstrates the difference in the convergence rate of the sequence $\{Q_t\}$ at different values of the revaluation coefficient γ.

Thus, the simulation modeling of the traffic light control process was carried out according the following conditions: the parameter of the revaluation coefficient is empirically selected and is equal to 0.1; the control decision is made every 8 s.

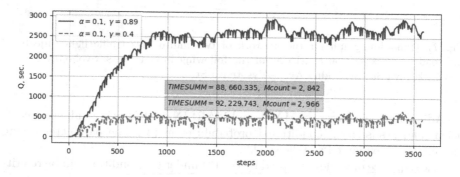

Fig. 9. Comparison of Q_t function at different values of γ.

3.3 FLCONTROL Module

In the FLCONTROL module, the output variable (duration of the traffic light phase) is calculated based on the defined rules (Table 2). The defuzzification is carried out using the centroid method. Figure 10 shows an example illustrating the decision to choose the duration of the phases of the next cycle at the time t. The calculated membership function of the fuzzy variable "phase" is marked in red for the value "medium", green for "long", blue for "short". The space under the graph is filled in one of the corresponding color depending on the decision to prolong or shorten the duration of the phase. The control result of the fuzzy controller is to give priority at a certain time of the day to directions with the most intensity traffic (Fig. 10).

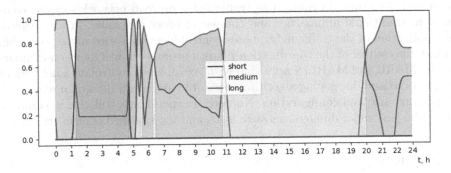

Fig. 10. Dynamics of changes in the values of membership functions of the terms of the output variable during the day. (Color figure online)

3.4   Model Comparison

A comparison of various efficiency indicators calculated for coordinated adaptive control of traffic light objects of the road network section (MARLIN), uncoordinated adaptive control (MARL), for traffic light objects with a fixed coordination plan (FIXED) and with variable length phases (FUZZY) is presented in Table 3. The MARL, MARLIN models outperform FIXED and FUZZY in all performance measures, and they are equally effective in the absence of unacceptable joint solutions.

Table 3. Effectiveness indicators comparison of joint control for different models.

Model	Average delay, $\left(\frac{sec.}{car}\right)$	Throughput, (car)	Delay, $(sec.)$
FIXED	14.9	4409	65770
MARL	9.4	4415	41360
Growth	5.5	6	24410
MARLIN	9.4	4412	41286
Growth	5.6	3	24484.2
FUZZY	14.9	4401	65668
Growth	0.0	−8	−102.8
MARL vs. FIXED	37.2%	0.1%	37.1%
MARLIN vs. FIXED	37.3%	0.1%	37.2%
MARL vs. FUZZY	37.2%	0.3%	37.0%

4   Conclusion

The conducted numerical experiments confirm the advantage of multi-agent app-roach and Q-learning in controlling traffic lights on road network sections com-pared to traditional approaches and the use of fuzzy controller, in this regard, the application of the traffic light phases control ideology seems more promising against the control of the coordination plans. Future work will focus on compar-ing the MARL and MARLIN approach with coordinated control on a network of intersections and developing algorithms for automatic agent design for a network of arbitrary size and configuration. Numerical experiments will aim to compare the traffic jam under different network loads and the specified control strategies.

References

1. Zadorozhny, V.N., Dolgushin, D., Yudin, E.B.: Analytical and Simulation Methods for Solving Actual Problems of System Analysis of Large Networks: A Monograph. OmSTU Publishing House, Omsk (2016)
2. Shvetsov, V.I.: Mathematical modeling of traffic flows. Avtomatika Telemechanika **11**, 3–46 (2003)
3. The AnyLogic Company official cite. https://www.anylogic.com. Accessed 12 Aug 2023
4. PTV VISUM official cite. https://www.ptvgroup.com. Accessed 12 Aug 2023
5. Charypar, D., Axhausen, K., Nagel, K.: Event-driven queue-based traffic flow microsimulation. Transp. Res. Rec. 35–40 (2003)
6. Note, T.: IEEE Trans. Syst. Sci. Cybern. **4**(4), 446–447 (1968). https://doi.org/10.1109/TSSC.1968.300174
7. Albrecht, S.V., Christianos, F., Schäfer, L.: Multi-agent Reinforcement Learning: Foundations and Modern Approaches. MIT Press, Cambridge (2023)
8. Treiber, M., Hennecke, A., Helbing, D.: Congested traffic states in empirical obser-vations and microscopic simulations. Phys. Rev. E **62**, 1805–1824 (2000)
9. Shoufeng, L., Ximin, L., Shiqiang, D.: Q-learning for adaptive traffic signal con-trol based on delay minimization strategy. In: Proceedings of IEEE Intetnational Conference on Networking, Sensing and Control, pp. 687–691 (2008). https://doi.org/10.1109/ICNSC.2008.4525304
10. El-Tantawy, S., Abdulhai, B., Abdelgawad, H.: Multiagent reinforcement learn-ing for integrated network of adaptive traffic signal controllers (MARLIN-ATSC): methodology and large-scale application on downtown Toronto. IEEE Trans. Intell. Transp. Syst. **14**(3), 1140–1150 (2013). https://doi.org/10.1109/TITS.2013.2255286
11. El-Tantawy, S., Abdulhai, B.: Towards multi-agent reinforcement learning for inte-grated network of optimal traffic controllers (MARLIN-OTC). Transp. Lett. Int. J. Transp. Res. **2**, 89–110 (2010). https://doi.org/10.3328/TL.2010.02.02.89-110
12. Wang, X., Sandholm, T.: Reinforcement learning to play an optimal Nash equilib-rium in team Markov games. In: Proceedings of the Advances Neural Information Processing Systems, p. 1571 (2002). https://doi.org/10.5555/2968618.2968817
13. Fukumoto, K., Fukumoto, O.: Multi-agent reinforcement learning: amodular app-roach. In: Proceedings of 2nd International Conference on Multi-agent Systems, pp. 252–258 (1996)

14. Chen, G., Cao, W., Chen, X., Wu, M.: Multi-agent Q-learning with joint state value approximation. In: Proceedings of the 30th Chinese Control Conference, Yantai, China (2011)
15. Komsiyah, S., Desvania, E.: Traffic lights analysis and simulation using fuzzy inference system of Mamdani on three-signaled intersections. Procedia Comput. Sci. **179**, 268–280 (2021). https://doi.org/10.1016/j.procs.2021.01.006
16. Tunc, I., Yesilyurt, A.Y., Soylemez, M.T.: Different Fuzzy Logic Control Strategies for Traffic Signal Timing Control with State Inputs. IFAC-PapersOnLine **54**, 265–270 (2021)
17. Teixeira, C.A., Villarreal, E.R.L., Cintra, M.E., Lima, N.W.B.: Proposal of a fuzzy control system for the management of traffic lights. IFAC Proc. Vol. **46**, 456–461 (2013). https://doi.org/10.3182/20130522-3-BR-4036.00062
18. Mine, H., Osaki, S.: Markovian Decision Processes, pp. 9–18. American Elsevier Publications Co., New York (1970)
19. Gasnikov, A.V., Gorbunov, E.A., Gooz, S.A.: Lektsii po sluchainym protsessam. Uchebnoe posobie, MIPT, Moscow, p. 285
20. Tislenko, T.I. , Semenova, D.V. , Sergeeva, N.A., Goldenok, E.E., Kononova, N.V.: Multiagent reinforcement learning for integrated network: applying to a part of the road network of krasnoyarsk city. In: IEEE 16th International Conference on Application of Information and Communication Technologies (AICT), pp. 1–5, Washington DC, USA (2022). https://doi.org/10.1109/AICT55583.2022.10013610
21. Tislenko, T.I. : The MARL task for the traffic light at the intersection. In: Materials of the VIII International Youth Scientific Conference "Mathematical and Software Support of Information, Technical and Economic Systems", pp. 144–149 (2021)
22. Tislenko, T.I. , Semenova, D.V. : The MARL task for the traffic light network. In: Information Technologies and Mathematical Modeling (ITMM-2021): Materials of the XX International Conference named after A.F. Terpugov. Tomsk (2022)
23. Tislenko, T.I. , Semenova, D.V. , Sergeeva, N.A.: Optimization of coordination plans for traffic light objects of the road network section. In: Control Systems, Information Technologies and Mathematical Modeling: Materials of the IV All-Russian Scientific and Practical Conference with International Participation (Omsk, 19 May 2022): in 2 v. / Ministry of Education and Science of Russia, Om. state tech. un-t, Dep. MM&ITE, pp. 255–260. Publishing House of OmSTU, Omsk (2022)
24. Zadeh, L.A.: Fuzzy Sets, Fuzzy Logic, Fuzzy Systems. World Scientific Press, Singapore (1996)
25. Sutton, R.S., Barto, A.G.: Reinforcement Learning: An Introduction, 2d edn, pp. 145–147. The MIT Press, Cambridge, London (2015)

Examining the Performance of a Distributed System Through the Application of Queuing Theory

Aleksandr Sokolov$^{(\boxtimes)}$ (ID), Olga Semenova (ID), and Andrey Larionov (ID)

V.A. Trapeznikov Institute of Control Sciences Russian Academy of Sciences,
Profsoyuznaya Street 65, 117997 Moscow, Russia
aleksandr.sokolov@phystech.edu

Abstract. The paper studies the performance of a distributed computing system using Markov chains and queuing theory. The system under study possesses a buffer capacity of N and M servers. Service time is distributed exponentially. Customers entering the system in a Poisson flow consist of a random number of tasks ranging from 1 to K. The probability that a customer contains k tasks is b_k, and the normalization condition is satisfied: $\sum_{k=1}^{K} b_k = 1$. Each individual task is serviced on a separate server. The order of service is determined by the FIFO principle. The paper describes a system using a Markov chain and provides formulas for calculating its performance characteristics. It compares the performance characteristics computed analytically to those obtained from a real distributed computing system.

Keywords: Queuing theory · Markov chain · Distributed computing

1 Introduction

To get accurate results when studying different problems, it's often necessary to do multiple calculations of the same type. This can involve running the same program with different input parameters. For example, in prediction tasks, data visualization, or machine learning tasks. The time required to obtain numerical results in tasks like these can range from a few minutes to weeks or even months, depending on the task's complexity. In order to speed up obtaining results, there are systems for distributed computing. These systems work by doing multiple tasks at the same time using several servers for task execution. Each task occupies a certain server and is executed on it. If M servers are provided in the system, then, accordingly, it is possible to execute M tasks simultaneously. The authors of this paper used a distributed computing system [11] to generate a synthetic dataset for the study of a complex prioritized system [14]. Using such tools, it is possible to reduce the time to obtain numerical results by tens of times.

The research was funded by the Russian Science Foundation, project no. 22-49-02023.

V. M. Vishnevskiy et al. (Eds.): DCCN 2023, CCIS 2129, pp. 16–32, 2024.
https://doi.org/10.1007/978-3-031-61835-2_2

There are a large number of systems used for scientific computations locally without providing capacity for public use. The Everest [13] and Nimrod [1] systems are examples. One special thing about the Nimrod system is that it has a language for describing tasks and can connect tasks together, using the output of one task as the input for another. The articles [5,7] consider a system for computing labor-intensive tasks JINR, in which OpenNebula technology is used for resource management. The main advantage of OpenNebula technology is that it allows you to combine virtual machines that use different virtualization technologies (such as VMware, KVM, LXD, and Firecracker) into a single computing cluster. Execution of computations on virtual machines is also realized in the Everest system [13]. Many local network parallel computation systems use the MPI library [9,12,15]. Numerous systems exist for organizing distributed computing. There are different types of computing systems. Some involve many users providing computing power (volunteer computing) [2–4,10], while others involve local distributed computing [5,7].

The analytical model of a class-based stream computing system is a system with group arrivals. Previous research has already delved into analytic models involving group arrivals. For example, in the paper [6] the $M[b]/M/1$ system is investigated, where groups with fixed size arrive, the system has only one server. In the work of [8] the $M[b]/M/N$ type of queuing systems is investigated, where group customer of fixed size are also received, but N servers are provided.

This paper presents a mathematical model for a distributed computing system. It assumes that the flow of customers follows a Poisson process with intensity λ, the service time is exponentially distributed with rate μ, and the number of tasks in a customer varies randomly according to a certain distribution. The probability that a customer consists of exactly k tasks is denoted by b_k, and the sum of all these probabilities from $k = 1$ to K (where K is the maximum number of tasks in a customer) is represented as $\sum_{k=1}^{K} b_k = 1$.

The paper is organized as follows. An example of a distributed computing system is provided in the "System for distributed computing" section. It explains the system architecture and the communication protocols used between the system components. In the "Analytical model" section, a Markov chain describing the distributed computing system is given. Formulas for calculating the stationary-state performance characteristics of the system are also given. In the "Numerical results" section, we study the performance characteristics of the system, such as the intensity of incoming flow, the average number of tasks in a customer, and the average time of service of one task. The characteristics calculated by analytical method are compared with the results obtained from the real system.

2   System for Distributed Computing

The system is a web application designed for organizing distributed computing. The application consists of several services: Backend-server, Supervisor and Worker, as well as the client part. The components of the system are shown in Fig. 1. The backend server offers an API for essential task operations such as launching, downloading results, and viewing statistics. The supervisor is responsible for efficiently managing and organizing the execution of queues of tasks. It plays a vital role in coordinating the queue of task execution among workers. The service effectively oversees task queues, monitors worker status, and logs outcomes in the database. A worker is a platform where tasks are carried out. You have the option to deploy this service on a dedicated virtual machine. The supervisor communicates with the workers using the HTTP protocol. Tasks input parameters and execution results are stored in a PostgreSQL database. The task queue is organized using a Redis in-memory database.

To ensure seamless interaction with the system, tasks are executed in Docker containers, regardless of the programming language or technologies being utilized.

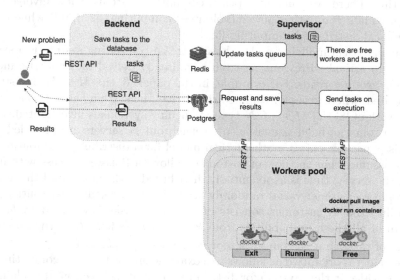

Fig. 1. System architecture for distributed computing.

In order to execute a task on the system, the user must generate a Dockerfile and construct a Docker image. Furthermore, it is necessary to upload the image to a publicly accessible repository like Docker Hub. To successfully build the image, the user is required to comply with two conventions. First, the input parameters inside the Docker container are passed into the environment variable **TASKS_PARAMS** as a string. Second, the results of executing a task in the docker container must be written to the **/tmp/results.json** file. To get the results after a task, there is a second convention. Users upload tasks to the system by providing a Docker image tag and attaching a JSON file with input parameters. Each array in the JSON file is a customer, and each separate element in the array is a task.

After the Docker image is uploaded, the customer is decomposed into separate tasks on the server, waiting to be executed. The supervisor watches for new tasks in the system and sends tasks to the queue for execution based on prioritization rules. As the workers become free, the supervisor sends the next highest priority task for execution. An application is considered completed if all tasks of this application are completed.

3   Analytical Model

In this section, we will delve into the analytical model of the distributed computing system. The system input (Fig. 2) receives customers with intensity λ, consisting of several tasks. We denote the number of tasks in the customer by $k = 1, .., K$, where K is some finite integer. Let us denote by b_k the probability that a customer contains k tasks to be serviced, $\sum_{k=1}^{K} b_k = 1$. The system consists of a buffer of capacity N (the maximum number of customers that can be placed in the buffer), where there are customers awaiting service and M servers. The service intensity is denoted by μ. In addition, for the sake of generality, we will assume that $K > M$, i.e., the maximum number of tasks in a customer is greater than the number of servers.

To build a Markov chain, we define that a customer is either being serviced or partially executed when at least one of its tasks is being serviced and there are other tasks of a customer waiting to be serviced. A customer is considered in the buffer or waiting for service when all its tasks are waiting for service. The state in the Markov chain, which describes a system for distributed computing, is represented by three numbers:

- n - number of customers in buffer;
- r - number of tasks completed of partially executed customers;
- m - number of occupied servers.

The states transition can be explained by a continuous-time Markov chain ξ_t.

$$\xi_t = \{(n, r, m), t \geq 0\} \tag{1}$$

with state space:

$$\Omega = \{(n, r, m), m \leq M, n = 0, r = 0 \cup$$
$$\cup (n, r, m), m = M, 0 \leq n \leq N, 0 \leq r \leq K\} \tag{2}$$

To understand Markov chain transitions for a system, it is helpful to break it into three fragments, describe them, and create balance equations for each state in each case. These fragments:

- $n = 0, r = 0, i \leq M$, the tasks of all customers are being serviced, there are no customers in the buffer and there are no tasks awaiting execution;

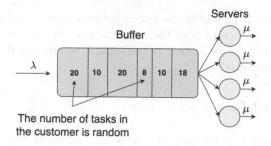

Fig. 2. System for distributed computations schema.

- $n = 0$, $r > 0$, $M + 1 \leq i \leq M + K$, when the system has no customers in the buffer, all servers are busy, but there are tasks of partially executed problems waiting to be executed;
- $n > 0$, there are customers in the buffer.

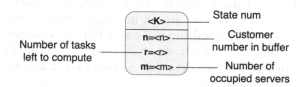

Fig. 3. State of the system in a Markov chain.

Figure 3 shows a schematic representation of the state of a Markov chain for a given system. For the probability of being in a certain state, it is convenient to introduce the notation $p(n, r, m)$.

3.1 Composition of the System of Balance Equations for the Case $n = 0$, $r = 0$, $i \leq M$, All Tasks in the System are Executed

We will now describe the system states when there are no customers in the buffer and no tasks waiting for execution. All tasks of all incoming customers are promptly sent for processing once they enter the system. The probability of being in the state when there are no customers in the system is equal to $p(0, 0, 0)$. Then the balance equation for the first state is:

$$\lambda p(0, 0, 0) = \mu p(0, 0, 1) \tag{3}$$

- the intensity of exit from a state is equal to the intensity of transition to this state. Figure 4 shows a diagram of a Markov chain. For the other state of this fragment, let us write out the balance equations:

$$(\lambda + i\mu)p(0,0,i) = (i+1)\mu p(0,0,i+1) + \sum_{j=0}^{i-1} \lambda b_{i-j}p(0,0,j),$$ (4)

$$1 \le i < M$$

Case $i = M$:

$$(\lambda + M\mu)p(0,0,M) = M\mu p(0,1,M) + \sum_{j=0}^{M-1} \lambda b_{M-j}p(0,0,j)$$ (5)

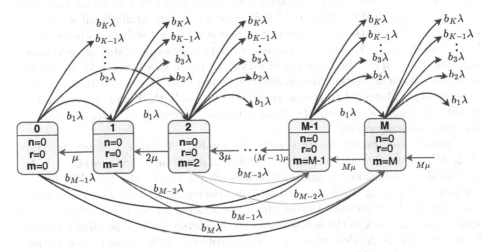

Fig. 4. Fragment of the Markov Chain for the first case.

Figure 4 shows a fragment of a Markov chain. Since $K > M$, there are possible transitions from all states to states $m = M$ and $r > 0$, i.e. not all tasks are immediately sent for service. It can be seen that for this fragment of the Markov chain, $M + 1$ equations are given. Rewrite the equations in a different form and number the states to create a system of equations for finding stationary states (Fig. 5).

3.2 Composition of the System of Balance Equations for a Fragment of Markov Chain, in the Case When There are Partially Executed Customers in the System and There Are No Customers in the Buffer $n = 0$, $r > 0$, $m = M$, $M + 1 \le i \le M + K$

This section explains a Markov chain fragment for a system where all servers are busy and there are partially served customers. There are no remaining customers in this situation. Figure 6a illustrates the transitions of this specific fragment of

	0	1	2	...	$i-1$	i	$i+1$	...	$M-1$	M	$M+1$
0	$-\lambda$	μ	0	...	0	0	0	...	0	0	0
1	λb_1	$-\lambda-\mu$	2μ	...	0	0	0	...	0	0	0
2	λb_2	λb_1	$-\lambda-2\mu$	...	0	0	0	...	0	0	0
⋮	⋮	⋮	⋮	...	⋮		...		⋮	⋮	⋮
i	λb_i	λb_{i-1}	λb_{i-2}	...	λb_1	$-\lambda-i\mu$	$(i+1)\mu$	...	0	0	0
$i+1$	λb_{i+1}	λb_i	λb_{i-1}	...	λb_2	λb_1	$-\lambda-(i+1)\mu$	...	0	0	0
⋮	⋮	⋮	⋮		⋮				⋮	⋮	⋮
$M-1$	λb_{M-1}	λb_{M-2}	λb_{M-3}	...	λb_{M-i}	λb_{M-1-i}	λb_{M-2-i}	...	$-\lambda-(M-1)\mu$	$M\mu$	0
M	λb_M	λb_{M-1}	λb_{M-2}	...	λb_{M-i+1}	λb_{M-1}	λb_{M-i-1}	...	λb_1	$-\lambda-M\mu$	$M\mu$

Fig. 5. First fragment: system of balances equations.

the chain. The states shown are from $M+1$-state, where all servers are busy and one customer is in the buffer, to $M+K+1$-state, where similarly all servers are busy, one task is in the queue, and there is one customer waiting to be serviced. In states $M+1, ..., M+K$ transitions between neighboring states are performed with intensity $M\mu$ - the intensity of service of the task when all M servers are busy, and also a transition from state $(0, r, M)$ to state $(1, r, M)$ with intensity λ is performed. The pending customer remains stored in the buffer. The transition to this state can occur from the states $(0, 0, m)$, where $m \le M$, as described in the previous subsection.

Consider separately the state with the number $M+K+1$. From this state $(1, 1, M)$ a transition with intensity λ (arrival of a customer) to the state $(2, 1, M)$ is possible. In addition, transitions to states with numbers $M+1, ..., M+K$ with intensities $M\mu b_1, ..., M\mu b_K$ are possible, respectively, depending on how many tasks are in the customer. During these transitions, a task is processed on the server, the only task in the buffer moves to the server, and the pending customer from the buffer is divided into k tasks with probabilities b_k. About the equations for this part of the Markov chain:

$$(\lambda + M\mu)p(0, r, M) = \mu M p(0, r+1, M) + \mu M b_r p(1, 1, M)$$

$$+ \sum_{j=0}^{M} \lambda b_{M-j+r} p(0, 0, j), \qquad (6)$$

$$1 \le r \le K-1$$

The left side of the equations represents the output intensity of the states. The first $\lambda p(0, r, M)$ is the transition intensity from the current state to the state $(1, r, M)$ (arrival of a new customer). The second $M\mu p(0, r, M)$ is the transition intensity to the state $(0, r-1, M)$ (servicing of the task with intensity $M\mu$). The right-hand side is the input intensity, which consists of three summands. The first $\mu M p(0, r+1, M)$ is the transition from the neighboring state $(0, r+1, M)$ to the current state $(0, r+1, M)$ with intensity μM (servicing the next task). The second $\mu M b_r p(1, 1, M)$ is the contribution of the transition from the state $(1, 1, M)$ when the tasks in the buffer run out and the customer is unpacked. The $\sum_{j=0}^{M} \lambda b_{M-j+r} p(0, 0, j)$ is the contribution of all possible transitions from the first fragment of the chain described above when not all attendants are busy.

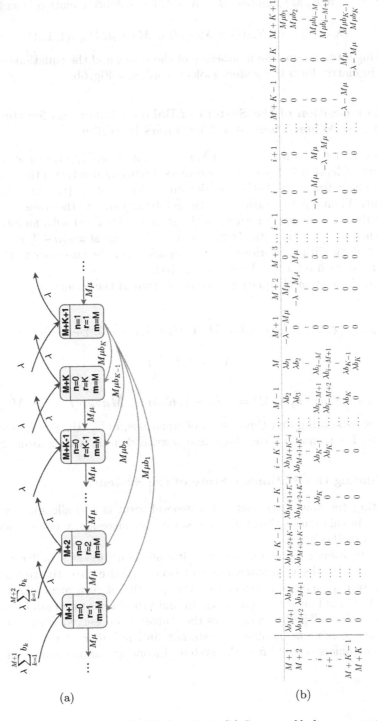

Fig. 6. (a) Second fragment of the Markov chain (b) System of balances equations for second fragment of Markov chain.

For state $(0, K, M)$ number $M + K + 1$, the balance equation is as follows:

$$(\lambda + M\mu)p(0, K, M) = \lambda b_K p(0, 0, M) + \mu M b_K p(1, 1, M) \tag{7}$$

Taking into account the numbering of the states and the equations compiled above, in matrix form the system looks accordingly Fig. 6b.

3.3   Composition of the System of Balance Equations for the Case $n > 0$, When There Are Customers in Buffer

This section details a fragment of a Markov chain describing transitions for systems with at least $n \geq 1$ pending customers. Figure 7a illustrates the transitions for this chain fragment. Let's consider an arbitrary state (n, r, m) within this fragment. When $r \neq 1$, transitions with an intensity of λ to the state $(n+1, r, m)$ are possible, as well as transitions to the state $(n, r - 1, m)$ with an intensity of $M\mu$. When $r = 1$, a transition with intensity λ to the state $(n + 1, r, m)$ is also possible, along with transitions during unpacking of the customer to the states $(n - 1, r, m)$ with intensity $M\mu b_r$, respectively.

Thus, the system of equations can be written in the form:

$$(\lambda + M\mu)p(n, r, M) = \lambda p(n - 1, r, M) + M\mu p(n, r + 1, M) + M\mu b_r p(n + 1, 1, M) \tag{8}$$

$$1 \leq r \leq K - 1$$

$$(\lambda + M\mu)p(n, K, M) = \lambda p(n - 1, K, M) + M\mu b_K p(n + 1, 1, M) \tag{9}$$

It is worth noting that this system of equations, unlike the previous sections, is infinite. Let us number the states and rewrite them in matrix form (Fig. 7b).

3.4   Finding the Stationary State of the System

Calculating the stationary state in a general form is a challenging endeavor. However, in this study, the stationary state was successfully determined using numerical methods.

To solve using a numerical method, it's necessary to set a buffer size limit. It is essential to derive a system of equations of finite size. To streamline the process, we implement a restriction on the quantity of customers allowed in the buffer. We should create a Markov chain and visualize the potential transitions within the chain. The Fig. 8 displays the transitions observed when the system reaches the maximum number of customers in the buffer. It can be seen that further customers do not enter the system. Incoming customers are lost.

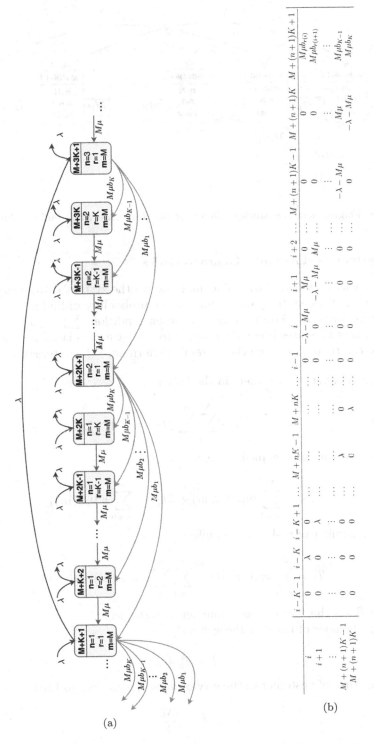

Fig. 7. (a) Third fragment of the Markov chain. (b) System of balances equations for third fragment of Markov chain.

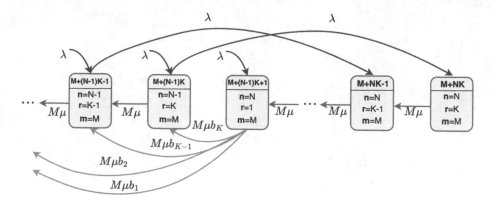

Fig. 8. Fragment of the Markov Chain for the third case when buffer is limited.

3.5 System Performance Characteristics

The equations provided above allow us to derive the system's stationary state. The solution of the system corresponds to the probability of finding the system in the state $p(n, r, m)$. From the normalization conditions $\sum_{n,r,m} p(n, r, m) = 1$ - the sum over all states is equal to one. Once we have calculated the stationary state, we can then determine the system's performance characteristics.

– average number of customers in the buffer:

$$\bar{N}_b = \sum_{n=1}^{N} \sum_{r=1}^{K} np(n, r, M) \tag{10}$$

– average number of occupied servers:

$$\bar{M} = \sum_{m=1}^{M} mp(0, 0, m) + M(1 - \sum_{m=0}^{M} p(0, 0, m)) \tag{11}$$

– average number of tasks in the buffer:

$$\bar{T}_b = \sum_{r=1}^{K} kp(0, r, M) + \sum_{n=1}^{N} \sum_{r=1}^{K} p(n, r, M)(n\bar{B} + r) \tag{12}$$

where $\bar{B}$ - is the average tasks number in customer
– average number of tasks in the system:

$$\bar{T} = \bar{T}_b + \bar{M} \tag{13}$$

– average time of customer in the service queue (according to Little's formula):

$$\bar{T}_p = \frac{N_b}{\lambda} \tag{14}$$

– the probability that the system is completely free:

$$P_{empty} = p(0,0,0) \tag{15}$$

– loss probability:

$$P_{loss} = \sum_{k=1}^{K} p(N, k, M) \tag{16}$$

4   Numerical Results

The analytical model was implemented in Python with the numpy library. In the first part, we studied how different system performance characteristics depend on system parameters. To validate the analytical model, we used the system described in the article [11]. Next, we described the input parameters and methods used to measure various metrics from the distributed computing system. All results are available in the https://github.com/timac11/multirunner-analytics.

4.1   Analysis of Performance Characteristics Through Analytical Methods

A study of distributed systems analyzed the following system characteristics: average number of customers in the buffer, number of busy servers, number of tasks in the buffer, number of tasks in the system and the probability of finding the system empty.

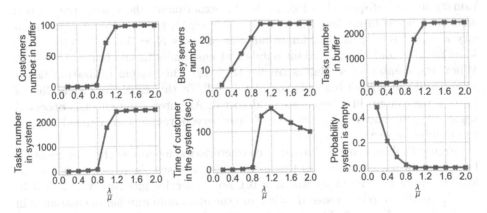

Fig. 9. Value of performance characteristics as a function of the ratio of intensity of incoming customers to service intensity.

Figure 9 shows how the system's performance characteristics depend on the value of $\frac{\lambda}{\mu}$. The parameters during the experiment were: the number of servers

$M = 25$, the maximum number of tasks $K = 50$, $b_k = 0.02$, $k = 1, ..., 50$, the maximum number of customers in the buffer $N = 100$. From the chart, it's evident that at $\frac{\lambda}{\mu} = 1.2$, the system reaches its limit, the buffer is full, and the system load factor ρ at $\frac{\lambda}{\mu} = 1.2$ $\rho \geq 1$. When $\frac{\lambda}{\mu} = 1.2$, the chance of the system being available approaches zero. When $\frac{\lambda}{\mu} \geq 1.2$, Little's formula cannot be used to calculate the average waiting time of a customer in the queue. Therefore, on the graph, when $\frac{\lambda}{\mu} \geq 1.2$, the waiting time decreases as the intensity increases, and the average number of customers in the buffer remains constant.

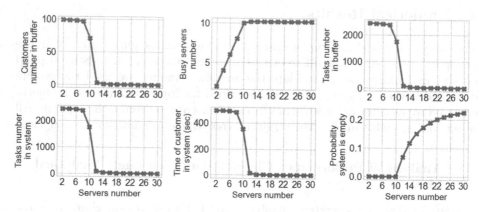

Fig. 10. Value of performance characteristics depending on the number of servers.

The dependence of performance characteristics on server's number in the stationary mode is displayed in Fig. 10. In this experiment, the system parameters were as follows: input flow intensity $\lambda = 0.5$ sec^{-1}, the task processing intensity $\mu = 0.2$ sec^{-1}, the maximum number of tasks $K = 50$, $b_k = 0.02$ where $k = 1, ..., 50$, and the maximum number of customers in the buffer $N = 100$. When the number of servers is small ($M \leq 10$), the system utilization factor $\rho \geq 1$ and the system buffer is completely full. In this case, the probability of finding the system in a free state is almost 0. An interesting effect is observed when $M \geq 10$: the average number of busy servers remains unchanged even as the number of servers increases.

Figure 11 shows how performance characteristics depend on the average number of tasks in a customer. In this experiment, the system parameters were as follows: input flow intensity was $\lambda = 0.1$ sec^{-1}, service intensity was $\mu = 0.15$ sec^{-1}, number of servers was $M = 20$, and the maximum number of customers in the buffer was $N = 100$. The average number of tasks per customer in the experiment ranged from 30 to 100. At an average number of tasks in the customer of $\hat{B} = 60$, the load factor in the system was $\rho \geq 1$.

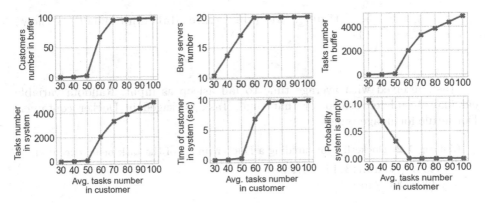

Fig. 11. Value of performance characteristics depending on the average number of tasks in the customer.

4.2 Comparison of Analytical Results with a Real System

The distributed computing system described earlier in the article was used to compare analytical calculations with metrics obtained from a real system. To simulate tasks whose execution times are distributed exponentially, a small Python program was written:

Listing 1.1. An example of a Python task in which the program falls asleep for some time distributed exponentially.

```python
from decouple import config
import time
import json
import numpy as np
import datetime


def sleep(num: int):
    time.sleep(np.random.exponential(num))
    print("Program sleeps {} seconds".format(num))
    return {"result": num}


with open("/tmp/result.json", "w") as writer:
    conf = config("TASK_PARAMS")
    num = int(json.loads(conf)["number"])
    now = datetime.datetime.now()
    res = sleep(num)
    writer.write(
        json.dumps(
            {
                "result": res,
```

$$"time": \ str(datetime.datetime.now() - now)$$
$$\}$$
$$)$$
$$)$$

Task received with average time to fall asleep as input. Random variable calculated and fell asleep for that time. Task packaged in Docker container. Dockerfile for image build:

Listing 1.2. Dockerfile for building a Docker image of the task being executed.

```
FROM python:3.11-slim
WORKDIR /usr

COPY requirements.txt .
COPY run.py .
RUN pip install -r requirements.txt --no-cache-dir

CMD [ "python", "./run.py" ]
```

Accordingly, the input file, in the case of service intensity $\mu = 0.1$, the number of tasks in the customer is 5, looks like this:

Listing 1.3. Example input file.

```
[
    {"number": 10},
    {"number": 10},
    {"number": 10},
    {"number": 10},
    {"number": 10}
]
```

We utilized a Python script and the system's REST API to upload files with intensity λ. In each experiment, we created approximately 50,000 customers to obtain statistical data.

The parameters were equal:

- $M = 15$ workers number;
- $\mu = 0.1 \ sec^{-1}$ entering customers intensity;
- $K = 100$ - max number of tasks in customer;
- $b_k = \frac{1}{100}$ for each $k = 1, 2, .., K - 1, K$.

The service intensity ranged from $\lambda = \frac{1}{100}$ to $\lambda = \frac{1}{20}$. Different aspects of the system's performance were studied depending on the value of λ. Experimental results (Fig. 12) show that the real system has a lower probability of catching the system free than the analytical results, possibly due to overhead caused by supervisor-worker interaction. The impact of asynchronous HTTP communication is described in detail in the paper [11].

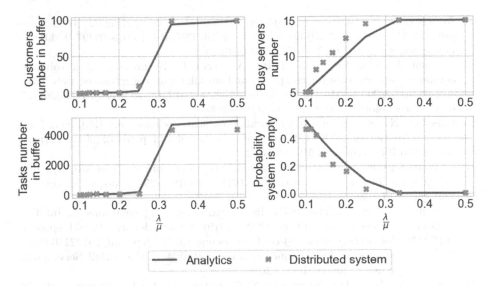

Fig. 12. Comparison of performance characteristics calculated by analytical method with values from the real system.

5 Conclusion

This paper presents an analytical model for exploring the performance of a distributed computing system. The model is built using Markov chains and queuing theory. This method calculates the stationary probabilities of the system. Evaluating important characteristics such as the average number of customers in the buffer, the average number of tasks in the system and buffer, and the average number of busy servers. Additionally, it analyzes system performance under varying parameters, such as flow intensity, average service time, and number of servers. The paper also includes a comparison between characteristics obtained from the analytical model and those observed in the real system.

References

1. Abramson, D., Giddy, J., Kotler, L.: High performance parametric modeling with Nimrod/G: killer application for the global grid? In: Proceedings 14th International Parallel and Distributed Processing Symposium. IPDPS 2000, pp. 520–528. IEEE Computer Society (2000). https://doi.org/10.1109/IPDPS.2000.846030, http://ieeexplore.ieee.org/document/846030/
2. Agliamzanov, R., Sit, M., Demir, I.: Hydrology@Home: a distributed volunteer computing framework for hydrological research and applications. J. Hydroinform. **22**(2), 235–248 (2020).https://doi.org/10.2166/hydro.2019.170, https://iwaponline.com/jh/article/22/2/235/71586/HydrologyHome-a-distributed-volunteer-computing

3. Anderson, D.P.: BOINC: a platform for volunteer computing. J. Grid Comput. **18**(1), 99–122 (2020).https://doi.org/10.1007/s10723-019-09497-9, http://link.springer.com/10.1007/s10723-019-09497-9

4. Antelmi, A., D'Ambrosio, G., Petta, A., Serra, L., Spagnuolo, C.: A volunteer computing architecture for computational workflows on decentralized web. IEEE Access **10**, 98993–99010 (2022). https://doi.org/10.1109/ACCESS.2022.3207167, https://ieeexplore.ieee.org/document/9893800/

5. Balashov, N., et al.: Service for parallel applications based on JINR cloud and HybriLIT resources. In: EPJ Web of Conferences, vol. 214, p. 07012 (2019). https://doi.org/10.1051/epjconf/201921407012

6. Ghimire, S., Ghimire, R.P., Thapa, G.B.: Mathematical models of Mb/M/1 bulk arrival queueing system. J. Inst. Eng. **10**(1), 184–191 (2014). https://doi.org/10.3126/jie.v10i1.10899

7. Korenkov, V., et al.: The JINR distributed computing environment. In: EPJ Web of Conferences, vol. 214, p. 03009 (2019). https://doi.org/10.1051/epjconf/201921403009, https://www.epj-conferences.org/10.1051/epjconf/201921403009

8. Kumar, J., Shinde, V.: Performance Evaluation Bulk Arrival and Bulk Service with Multi Server using Queue Model (2018)

9. Nguyen, N., Bein, D.: Distributed MPI cluster with Docker Swarm mode. In: 2017 IEEE 7th Annual Computing and Communication Workshop and Conference (CCWC), pp. 1–7. IEEE (2017). https://doi.org/10.1109/CCWC.2017.7868429, http://ieeexplore.ieee.org/document/7868429/

10. Nikitina, N., Manzyuk, M., Podlipnik, Č., Jukić, M.: Volunteer computing project SiDock@home for virtual drug screening against SARS-CoV-2, pp. 23–34 (2021). https://doi.org/10.1007/978-3-030-86582-5_3, https://link.springer.com/10.1007/978-3-030-86582-5_3

11. Sokolov, A., Larionov, A., Mukhtarov, A., Fedotov, I.: Architecture of a distributed parallel computing system using docker cluster. In: Proceedings of the 2022 International Conference on Information, Control, and Communication Technologies. ICCT 2022 (2022). https://doi.org/10.1109/ICCT56057.2022.9976525

12. Stelly, C., Roussev, V.: SCARF: a container-based approach to cloud-scale digital forensic processing. In: DFRWS 2017 USA - Proceedings of the 17th Annual DFRWS USA, pp. S39–S47 (2017). https://doi.org/10.1016/j.diin.2017.06.008

13. Sukhoroslov, O., Putilina, E.: Cloud services for automation of scientific and engineering computations science. Bus. Soc. **1**(2), 6–9 (2018)

14. Vishnevsky, V., Klimenok, V., Sokolov, A., Larionov, A.: Performance evaluation of the priority multi-server system mmap/ph/m/n using machine learning methods. Mathematics **9**(24) (2021).https://doi.org/10.3390/math9243236

15. Zhou, J., Bie, S.W., Miao, L., Zhang, Y., Jiang, J.: DOCKER-enabled scalable parallel mlfma system for RCS evaluation. Progr. Electromagn. Res. M **67**, 169–176 (2018). https://doi.org/10.2528/PIERM18021907, http://www.jpier.org/PIERM/pier.php?paper=18021907

Selecting the Performance Metrics to Control the CPU Oversubscription Ratio in a Cloud Server

Ruslan Smeliansky[1], Vasily Balashov[1]([⊠]), Vasily Pashkov[1],
Alexey Bergovin[1], and Margarita Orlova[2]

[1] Lomonosov Moscow State University, Leninskie Gory, MSU, 1, Bldg. 52,
119991 Moscow, Russia
{smel,hbd,pashkov}@cs.msu.ru
[2] Moscow Power Engineering Institute, Krasnokazarmennaya 14, Bldg. 1,
111250 Moscow, Russia
OrlovaMA@mpei.ru

Abstract. Virtual machines (VMs) in IaaS public clouds often request more resources than will actually be used, in particular the CPU resource expressed in the number of cores. To improve CPU utilization, the cloud server performs CPU oversubscription, i.e. offers more cores to VMs than it actually has. To avoid degradation of user applications performance (e.g. significant increase of application latency) due to oversubscription, it is necessary to monitor server performance metrics that correlate with application latency. In this paper, we address the problem of selecting such metrics. Selection is performed on the base of experiments with synthetic workload on a developed testbed, which includes VMs with preconfigured applications, VMs with request generators, and servers to run these two sets of VMs separately. Over 120 metrics are collected on the server running VMs with applications, and in every VM with application. Several series of experiments were performed with homogeneous and heterogeneous workloads. For every series of experiments, a metric best correlating with application latency was selected. Finally, a candidate for "universal" metric was proposed, based on the host runqueue length. The experimentally selected metrics can be used in a group of methods that control CPU oversubscription ratio based on monitoring of performance metrics.

Keywords: IaaS public cloud · virtual machine · CPU oversubscription · performance metric

1 Introduction

In Infrastructure-as-a-Service (IaaS) public clouds, servers run user-supplied virtual machines (VMs) with applications not under control of the cloud provider. The user requests a certain amount of virtual resources (CPU cores, memory,

V. M. Vishnevskiy et al. (Eds.): DCCN 2023, CCIS 2129, pp. 33–45, 2024.
https://doi.org/10.1007/978-3-031-61835-2_3

disk storage) for the VM. It is often the case that the requested amount of resources is significantly underused. Therefore if physical resources of the server are allocated to the VM in the same amount as requested (1:1 ratio), the server is underloaded. Resource oversubscription is used to deal with this problem, meaning that the server offers more resources to the VMs than it is actually available. In this paper we focus on CPU oversubscription (OVS, for short). For instance, 3:1 CPU OVS ratio for a server with 8 CPU cores means that a set of VMs with up to 24 virtual CPU cores can be run on the server. For sake of brevity, we will refer to physical CPU cores as pCPU, and to virtual CPU cores as vCPU.

Best-practice recommendations [1] suggest up to 2:1 OVS ratio for business-critical or latency-sensitive applications, and 2:1 to 4:1 OVS ratio for non-critical or latency-insensitive applications. We will consider the case in which the user does not indicate in the service level agreement (SLA) that the application in the VM is business-critical. In such case it is reasonable to reach OVS ratios higher than 2:1, but with high OVS ratio there is a risk of significant application performance degradation in case the VMs exhibit high demand for CPU resource. For example, if a user application is servicing external requests at the rate that creates 20% CPU load without expressed peaks, then 3:1 OVS ratio is acceptable, but for more intense workloads there is a need for detailed monitoring of server performance to choose the OVS ratio.

There is a substantial amount of research on how to predict the actual demand of a VM for CPU resource based on VM execution in a controlled environment with monitoring of CPU-oriented performance metrics. In [2] such metric is the CPU scheduling (allocation) latency, i.e. the time a ready process must wait for CPU. The authors of [3] suggest monitoring the CPU load, which is a somewhat trivial metric. Predictive methods in [4] operate on Kubernetes containers and consider the "CPU slack" metric, which is the difference between the amount of requested vCPUs and actual CPU load by the container. In [5], the metric is CPU load per gigabyte of network traffic; such metric can be confusing, as it may reach a high value in case of moderately CPU-intensive calculations with small input data. A multiple constraint multiple knapsack problem considered in [6] is focused on the VM's CPU runqueue length, which is interrelated with CPU scheduling latency.

With such variety of metrics used in the existing methods, it is reasonable to analyze which metric(s) best correlate with the user application performance. Such metrics can be recommended for use in metrics-oriented methods for CPU usage prediction aimed at choosing the CPU OVS ratio. In this paper we perform such analysis, taking application response time (or latency) as the measure of its performance, which is not directly observable by the public cloud provider – thus the need to monitor the metrics. A single server is considered, and it is assumed that VMs on the server are independent, i.e. do not communicate with each other.

The rest of the paper is structured as follows. Section 2 describes the set of applications for synthetic workload. In Sect. 3, the set of collected performance metrics is presented with tools for their collection. Section 4 describes the organization of the experimental testbed, including its structure and usage procedure. Section 5 presents the performed experiments, along with resulting selection of metrics that best correlate with applications latency. In Sect. 6 the final remarks are made, including the suggestions on applying the obtained results.

2   Selection of Applications for the Synthetic Workload

For simulation of server workload in the experiments, following workload sources were considered:

1. Synthetic cloud benchmarks: SPEC Cloud IaaS [7], CloudSuite [8], Tail-Bench [9].
2. Workload generators for benchmarking real applications:
    - for web servers: wrk [10], Apache JMeter [11], ApacheBench [12], Siege [13];
    - for database management systems: Yahoo! Cloud Serving Benchmarking (YCSB) [14], TPC-H [15], TPCC-MySQL [16];
3. Deep Learning applications for image recognition with a subset of ImageNet dataset [17] as input: AlexNet [18], LeNet [19], ResNet [20].

These workload sources create steady streams of requests to applications. This corresponds to the static workload pattern. Typical patterns of cloud workload, such as static, periodic, on-and-off, and unpredictable (irregular) are described in [21]. The analysis of AliBaba 2017–2018 set of cluster traces performed in [22] shows that CPU utilization (load) for servers running latency-sensitive "service" applications rarely reaches 15% (so there is plenty of room for CPU oversubscription), and these applications have mostly static workload pattern. This justifies our selection of workload sources for creating the synthetic workload.

Final selection of workload sources is presented in Table 1. Real applications with separate generators of requests were preferred over synthetic benchmarks, because the former case requires less specialization of VMs with applications, and the separate request generators allow more control over intensity of requests than in synthetic benchmarks. AlexNet was chosen because of more mature image recognition techniques than in the two remaining Deep Learning applications. "Empty" VM, with no application besides SSH server for remote access, was included to simulate inactive VMs which constitute a significant subset of VMs in a public cloud [23].

Table 1. Workload sources for experiments.

Workload type	Application	Benchmark/ Request generator
Empty VM	None	None
Memory-intensive	Redis in-memory datastore	YCSB
Disk I/O-intensive	MySql database	TPCC-mysql
CPU-intensive	AlexNet image recognition	Subset of ImageNet
General (mixed resource usage)	Apache web server+WordPress CMS	JMeter

In the testbed, every instance of an application and every instance of a requests generator runs in a dedicated VM. VMs with applications and VMs with request generators run on separate servers.

3   The Set of Collected Performance Metrics

To identify the performance metrics that best correlate with application latency, we collected an extensive set of metrics both for the host (server) and for every VM running on it. The metrics for different types of resources are listed in Table 2, along with the tools used for collection of these metrics. These tools are available in Debian 10 64-bit OS used in the testbed for both the host and the VMs. The tools are commonly available in modern Linux distributions. Detailed meaning of the metrics can be found in the documentation on the metric collection tools; several key metrics are explained below. Note that in IaaS public clouds it is more reasonable to use host-based metrics than VM-based metrics, as the VMs are often outside the cloud provider's control, and metrics from them may be unavailable.

Over 120 metrics were collected both on the host and in every VM. Values of the metrics were saved every 10 s. In order to decrease metrics monitoring overhead, in the production server only the metrics that best correlate with application latency should be monitored.

Based on literature overview and the authors' experience, the original assumption was that the following metrics are candidates for high correlation with application latency:

1. CPU runqueue length: the number of processes waiting for CPU time;
2. CPU steal time [24]: the percentage of time a virtual CPU waits for physical CPU time (because the hypervisor is servicing another VM);
3. CPU pressure [25]: the percentage of time in which at least some processes are stalled waiting for CPU resource.

The difference between CPU steal time and CPU pressure is that CPU pressure can be high even if a vCPU almost always has pCPU resource (e.g. one VM

Table 2. Collected performance metrics.

Tool	Metric names
Metrics for CPU as a whole	
vmstat	procs_r, procs_b, cpu_us, cpu_sy, cpu_id, cpu_wa, cpu_st, system_in, system_cs
/proc/pressure/cpu	cpu_some_avg10, cpu_some_avg60, cpu_some_avg300, cpu_some_total
uptime	cpu_load_1, cpu_load_5, cpu_load_15
/proc/cpuinfo	proc_cpu_MHz
Metrics for individual CPU cores	
/proc/softirqs	N_softirq
/proc/interrupts	N_hardirq
mpstat	usr, nice, sys, iowait, irq, soft, guest, gnice, idle, steal
Metrics for memory	
vmstat	memory_swpd, memory_free, memory_buf, memory_cache, swap_si, swap_so
/proc/pressure/memory	memory_some_avg10, memory_some_avg60, memory_some_avg300, memory_some_total, memory_full_avg10, memory_full_avg60, memory_full_avg300, memory_full_total
numastat	MemTotal, MemFree, MemUsed, Active, Inactive, Active(anon), Inactive(anon), Active(file), Inactive(file), Unevictable, Mlocked, Dirty, Writeback
sar	sar_pgpgin, sar_pgpgout, sar_fault, sar_majflt, sar_pgfree, sar_pgscank, sar_pgscand, sar_pgsteal, sar_vmeff, sar_kbmemfree, sar_kbavail, sar_kbmemused, sar_memused, sar_kbbuffers, sar_kbcached, sar_kbcommit, sar_commit, sar_kbactive, sar_kbinact, sar_kbdirty, sar_kbanonpg, sar_kbslab, sar_kbkstack, sar_kbpgtbl, sar_kbvmused
Metrics for disk input/output	
vmstat	io_bi, io_bo
/proc/pressure/io	io_some_avg10, io_some_avg60, io_some_avg300, io_some_total, io_full_avg10, io_full_avg60, io_full_avg300, io_full_total
biosnoop	io_latency
sar	sar_bread, sar_lread, sar_rcache, sar_bwrit, sar_lwrite, sar_wcache, sar_rtps, sar_wtps, sar_tps, sar_dtps, sar_scall, sar_badcall
Metrics for network input/output	
netstat -s	IpExt_OutMcastPkts, IpExt_InBcastPkts, IpExt_InOctets, IpExt_OutOctets, IpExt_OutMcastOctets, IpExt_InBcastOctets, IpExt_InNoECTPkts, IP_rcv, IP_fwd, IP_dsc, IP_snt, IP_drp, IP_dlv
sar	sar_total, sar_dropd, sar_squeezd, sar_rx_rps, sar_flw_lim

per host), and thus CPU steal time is close to zero. For I/O-intensive applications, I/O pressure [25] (for processes stalled waiting for I/O resource) can be a candidate for high correlation with application latency.

4   Testbed Configuration and Usage

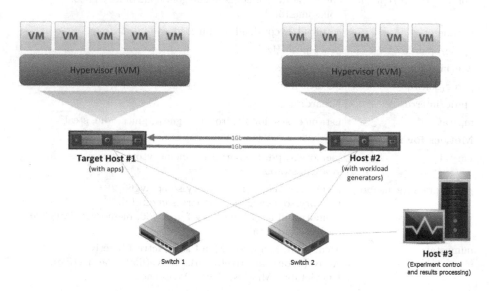

Fig. 1. Experimental testbed structure.

To perform experimental research of correlation between metrics values and applications latency, a testbed was developed. Its structure is shown in Fig. 1. The tesbed includes three hosts (servers):

1. host for running the VMs with applications;
2. host for running the VMs with workload generators;
3. host for experiments control and results processing.

All three hosts operate under Debian 10 64-bit OS. Hosts 1 and 2 use KVM hypervisor with QEMU virtual machine manager. VMs also use Debian 10 64-bit OS.

The most loaded host 1 has the following hardware configuration:

- CPU: Intel Xeon 3.2 GHz, 4 cores, 2 threads/core;
- RAM: 32 Gb;
- SSD: 800 Gb;
- network: Gigabit Ethernet.

Typical resources configuration for a VM is:

- one vCPU;
- RAM: 1 or 2 Gb;
- storage: 40 Gb;

The procedure for testbed usage in the experiment is as follows:

1. create the necessary number of VMs with applications (on host 1) and with workload generators (on host 2) by cloning preconfigured VM images available on these hosts;
2. start the VMs, wait for software initialization;
3. configure the virtual network connections for every pair <VM with application, VM with workload generator>;
4. start the workload generators and begin collection of metrics (on host 1 and on VMs with applications) and latency values (on VMs with workload generators);
5. wait for the duration of the active experiment phase (configurable, set to 60 min);
6. finish the collection of metrics and latencies, stop the workload generators;
7. copy to host 3 the log files with metrics values from host 1 and from every VM with application;
8. copy to host 3 the log files with applications latency values from host 2;
9. perform the experiment results processing and analysis:
   - logs conversion and merging to a CSV spreadsheet;
   - for every host-based metric: calculation of maximum, minimum, and average values;
   - for every VM-based metric: calculation of maximum, minimum, and average values; averaging the results between VMs, producing VM-averaged maximum, minimum and average;
   - for application latency of every VM with same application: processing in the same way as a VM-based metric.

This procedure is automated with scripts developed for the experimental research. Preliminary experiments indicated that the applications selected for the workload (see Table 1) have a warm-up phase of several minutes, characterized with highly unstable latencies (Fig. 2). Such effects do not significantly influence application performance in the long run, so these warm-up phases were excluded from statistical analysis to avoid bias in case of experiments limited to 1 h each.

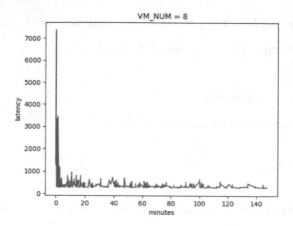

Fig. 2. Application latency of one VM with a warm-up phase, for 8 Apache+WordPress VMs on a host.

5   Experiments Methodology and Results

5.1   Experiments Goal and Setup

The goal of the experimental research is to identify the performance metrics that best correlate with application latency. It is reasonable to suppose that for different types of VMs constituting the workload (VM type is defined by its application) the leading metrics can be different. Therefore, we will perform several series of experiments with scaling up the number of VMs:

1. homogeneous workload: all VMs have the same type, number of VMs increases;
2. heterogeneous workload: the number of VMs of one type increases, the set of VMs with other types is fixed.

Since in our experiments every VM with application requests one vCPU, and the number of pCPUs on the server (host 1) is fixed, we will analyze correlation between the application latency and the number of VMs in the workload.

5.2   Procedure for Metric Selection

The procedure for metric selection in a series of experiments is as follows:

1. perform a series of experiments on the testbed with increasing number of VMs with one type (and, possibly, fixed set of VMs with other types); the results are following arrays, each with the number of elements equal to the number of experiments:
   a. for every host-based metric: maximum, minimum and average values (3 arrays);

    b. for every VM-based metric: maximum, minimum and average values (3 arrays);

    c. for application latency: average values (1 array).

2. for every array from the items a and b, calculate ρ – its Pearson correlation coefficient with array of latencies (item c);
3. *filter 1*: select the best metrics ($|\rho| \geq 0.7$), sort them by non-increase of $|\rho|$;
4. select one metric M from the top of the list;
5. for every metric selected in step 3, except M, calculate ρ_M – its Pearson correlation coefficient with M;
6. *filter 2*: remove the metrics highly ($|\rho_M| \geq 0.7$) correlating with M.

Averaging the application latency (for the same application within an experiment) is acceptable, as the experiments show that the latency peaks are located in the warm-up phase of VM operation, which is discarded in the experiment results analysis.

5.3 Experiment Results and Analysis

Following series of experiments were performed:

- homogeneous workload:
  - WordPress+Apache: 1 to 24 VMs;
  - MySQL: 1 to 12 VMs;
  - Redis: 1 to 24 VMs;
  - AlexNet: 1 to 24 VMs;
- heterogeneous workload:
  - fixed subset: 1 Redis, 1 AlexNet, 2 Empty;
  - variable subset: 2 to 12 WordPress+Apache (latency is analyzed for this type of VMs).

Best metrics (with largest absolute values of ρ) for different series of experiments are shown in Table 3. Average, minimum and maximum values are marked with "mean", "min" and "max" suffixes. Host-based metric names begin with "HST".

Based on Table 3, we select the following metrics for corresponding types of workload.

For WordPress+Apache, we select HST_cpu_some_avg10_mean, which is a host-based CPU pressure metric averaged for 10 s. Other metrics do not pass *filter 2*.

For MySQL, we select HST_vmstat_r_mean, because it is a host-based CPU-oriented metric, and its correlation difference with better (but VM-based and memory-oriented) metrics is less than 0.01. Other metrics do not pass *filter 2*.

For Redis, we select HST_vmstat_r_max, for sake of unification with the metric for heterogeneous workload (see below). This metric is a host-based metric indicating the maximum runqueue length, and its correlation difference with better metrics is less than 0.01. One metric passes *filter 2*, namely

Table 3. Best metrics for the series of experiments.

Metric	ρ	Metric	ρ
WordPress+Apache			
HST_cpu_some_avg10_mean	0.974223	HST_cpu_some_avg300_mean	0.974204
HST_cpu_some_avg60_mean	0.974004	HST_cpu_some_avg300_max	0.971051
HST_vmstat_sy_mean	0.954290	HST_uptime_cpu_load_1_max	0.950528
HST_cpu_some_avg60_max	0.936790	HST_cpu_some_total_max	0.932633
HST_uptime_cpu_load_5_max	0.906681	HST_vmstat_r_mean	0.905192
MySQL			
sar_mem_kbmemused_mean	-0.997691	sar_mem_%memused_mean	-0.997685
sar_mem_kbanonpg_mean	-0.995365	sar_mem_kbinact_mean	0.994469
mem_some_avg60_max	0.993676	sar_mem_kbactive_mean	-0.993210
mem_some_avg300_max	0.992691	sar_mem_kbbuffers_mean	0.992616
mem_some_avg10_max	0.992210	vmstat_buff_mean	0.991816
sar_mem_kbavail_mean	0.991233	sar_mem_kbmemfree_mean	0.991207
vmstat_free_mean	0.991177	sar_mem_kbvmused_mean	-0.991086
sar_mem_kbkstack_mean	-0.990101	HST_vmstat_r_mean	0.990057
Redis			
HST_uptime_cpu_load_15_mean	0.998196	HST_uptime_cpu_load_5_mean	0.997381
HST_uptime_cpu_load_15_max	0.996493	HST_uptime_cpu_load_1_mean	0.996338
HST_uptime_cpu_load_5_max	0.996149	HST_vmstat_r_mean	0.995390
HST_uptime_cpu_load_1_max	0.994016	HST_sar_mem_kbvmused_mean	0.990850
HST_vmstat_r_max	0.990699	HST_uptime_cpu_load_5_min	0.990561
AlexNet			
mpstat_%idle_mean	-0.999994	sar_mem_kbavail_mean	-0.999987
sar_mem_kbmemused_mean	0.999986	sar_mem_%memused_mean	0.999986
sar_mem_kbanonpg_mean	0.999975	sar_mem_kbpgtbl_mean	0.999962
sar_mem_kbcommit_mean	0.999962	sar_mem_%commit_mean	0.999962
uptime_cpu_load_5_mean	0.999961	uptime_cpu_load_1_mean	0.999950
uptime_cpu_load_15_mean	0.999929	vmstat_st_mean	0.999658
mpstat_%steal_mean	0.999617	cpu_some_total_mean	0.998565
HST_sar_mem_kbkstack_max	0.997716	HST_vmstat_bo_mean	0.997664
HST_vmstat_so_mean	0.997189	HST_vmstat_so_max	0.996978
sar_mem_kbinact_mean	0.996937	HST_cpu_some_avg10_mean	0.996841
HST_cpu_some_avg60_min	0.996695	HST_cpu_some_avg60_max	0.996661
HST_cpu_some_avg300_max	0.996453	HST_vmstat_cache_max	0.996379
HST_sar_bread/s_mean	0.996228	HST_vmstat_b_max	0.996130
HST_mpstat_%steal_mean	0.996078		
Heterogeneous			
HST_vmstat_r_max	0.987049	HST_sar_tps_max	0.972972
HST_mpstat_%iowait_mean	0.969527	HST_io_some_total_max	0.939969
HST_io_full_total_max	0.937409	HST_sar_mem_kbinact_mean	0.934054
HST_sar_mem_kbinact_max	0.931242	HST_vmstat_cs_max	0.930470
HST_io_some_total_mean	0.922245	HST_sar_rtps_max	0.919280

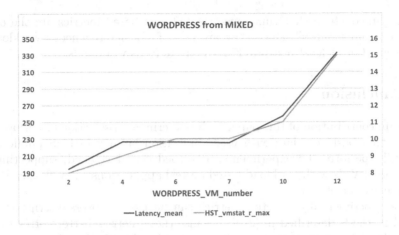

Fig. 3. Matching of WordPress+Apache latency (left axis) and the selected metric (right axis) for heterogeneous workload.

HST_io_some_avg60_min with $\rho_M = 0.63$. It has marginal correlation with latency ($\rho = 0.7$) and there is no reason to consider it a second metric.

For AlexNet, we select HST_cpu_some_avg10_mean. Other metrics do not pass *filter 2*.

For heterogeneous workload with growing number of WordPress+Apache VMs we see a clear difference in best metrics list to the list for pure Word-Press+Apache workload. This indicates that presence of VMs with other applications has its influence on the correlation of metrics with WordPress+Apache latency. The evident leader is HST_vmstat_r_max. Other metrics do not pass *filter 2*.

Figure 3 illustrates high correlation between the application latency and the value of the selected metric in case of heterogeneous workload.

The question remains – is it possible, judging by the performed experiments, to propose a "universal" metric, useful for both the explored homogeneous and heterogeneous workloads. Experiments with heterogeneous workload clearly suggest to monitor the runqueue length on the host. Let us estimate the relevance of runqueue-oriented host-based metrics for the considered types of homogeneous workload:

- Redis: HST_vmstat_r_max was already selected, with $\rho \approx 0.99$;
- MySQL: HST_vmstat_r_mean was already selected, with $\rho \approx 0.99$;
- AlexNet: HST_vmstat_r_mean has $\rho \approx 0.985$, and can be safely used instead of HST_cpu_some_avg10_mean;
- WordPress+Apache: HST_vmstat_r_mean has $\rho \approx 0.905$; while the correlation is high, this is the case in which CPU pressure-based metric is still preferable for homogeneous workload.

We can conclude that runqueue-oriented host-based metrics are the candidates for "universality", since they are good for a heterogeneous workload as well as good or at least acceptable for homogeneous workloads.

6   Conclusion

The main contribution of this paper is the experimental selection of performance metrics that best correlate with user application latency in IaaS public cloud server. To perform the experiments, a testbed was developed, supporting the execution of VMs with workload based on real applications, as well as collecting an extensive set of metrics.

The experimentally selected metrics can be used in oversubscription ratio control methods described in [2–4], because these metrics better correlate with application latency than those originally considered in the listed papers.

Another approach to use of the selected metrics is to find the boundary value of a metric, which corresponds to boundary value of the application's latency, and monitor the value of the metric instead of the value of application latency (which is unobservable by the cloud provider). This approach is more suitable for software-as-a-service (SaaS) clouds, in which the applications and their boundary latencies are known to the provider.

As all the selected metrics are host-based, they do not require instrumentation of user-controlled VMs, and therefore are usable in IaaS public clouds, where the servers are under provider's control. It should be noted that VMs colocated on a server must be continuously monitored for resource usage pattern and intensity, to detect possible changes of both factors, leading to the necessity of VM migration in order to avoid server overloads.

Acknowledgments. The authors express their gratitude towards Ariy Okonishnikov for his key role in performing the experiments on the testbed.

References

1. Nutanix Best Practices (2020). https://softprom.com/sites/default/files/materials/BP-2029_AHV_compressed.pdf. Accessed 30 Jan 2024
2. Bashir, N., Deng, N., Rzadca, K., Irwin, D., Kodak, S., Jnagal, R.: Take it to the limit: peak prediction-driven resource overcommitment in datacenters. In: Proceedings 16th European Conference on Computer Systems, pp. 556–573 (2021)
3. Langston, N., Krishnan, R.: Using LSTM and SARIMA Models to Forecast Cluster CPU Usage. arXiv preprint arXiv:2007.08092 (2020)
4. Wang, T., Ferlin, S., Chiesa, M.: Predicting CPU Usage for Proactive Autoscaling. In: Proceedings of 1st Workshop on Machine Learning and Systems, pp. 31–38 (2021)
5. Peng, H., Wen, W.S., Tseng, M.L., Li, L.L.: A cloud load forecasting model with nonlinear changes using whale optimization algorithm hybrid strategy. Soft. Comput. **25**(15), 10205–10220 (2021)

6. Baset, S.A., Wang, L., Tang, C.: Towards an understanding of oversubscription in cloud. In: Proceedings of 2nd USENIX Workshop on Hot Topics in Management of Internet, Cloud, and Enterprise Networks and Services (Hot-ICE 2012), 6p. (2012)
7. SPEC Cloud IaaS 2018. https://www.spec.org/cloud_iaas2018/. Accessed 30 Jan 2024
8. CloudSuite 4.0, https://www.cloudsuite.ch/, source code: https://github.com/parsa-epfl/cloudsuite. Accessed 30 Jan 2024
9. TailBench v0.9. http://tailbench.csail.mit.edu/, source code: https://github.com/supreethkurpad/Tailbench. Accessed 30 Jan 2024
10. wrk – a HTTP benchmarking tool, source code: https://github.com/wg/wrk. Accessed 30 Jan 2024
11. Apache JMeter 5.5. https://jmeter.apache.org/, source code: https://jmeter.apache.org/download_jmeter.cgi. Accessed 30 Jan 2024
12. ApacheBench 2.4.57. https://httpd.apache.org/docs/current/programs/ab.html, source code: https://httpd.apache.org/download.cgi. Accessed 30 Jan 2024
13. Siege 3.1.4. http://www.joedog.org/, source code: https://github.com/JoeDog/siege. Accessed 30 Jan 2024
14. Yahoo! Cloud Serving Benchmarking. https://github.com/brianfrankcooper/YCSB/wiki, source code: https://github.com/brianfrankcooper/YCSB. Accessed 30 Jan 2024
15. TPC-H 3.0.1. https://www.tpc.org/tpch/, source code: https://www.tpc.org/tpc_documents_current_versions/current_specifications5.asp. Accessed 30 Jan 2024
16. TPCC-MySQL, source code: https://github.com/Percona-Lab/tpcc-mysql. Accessed 30 Jan 2024
17. ImageNet image database. https://image-net.org/. Accessed 30 Jan 2024
18. AlexNet implementation in pytorch. https://github.com/pytorch/vision/blob/main/torchvision/models/alexnet.py. Accessed 30 Jan 2024
19. LeNet-5. https://github.com/activatedgeek/LeNet-5. Accessed 30 Jan 2024
20. ResNet implementation in pytorch. https://github.com/pytorch/vision/blob/main/torchvision/models/resnet.py. Accessed 30 Jan 2024
21. Nikravesh, A.Y., Ajila, S.A., Long, C.-H.: An autonomic prediction suite for cloud resource provisioning. J. Cloud Comput. Adv. Syst. App. **6**(3), 20 (2013)
22. Everman, B., Rajendran, N., Li, X., Zong, Z.: Improving the cost efficiency of large-scale cloud systems running hybrid workloads - a case study of Alibaba cluster traces. Sustain. Comput. Inform. Syst. **30**, 11 (2021)
23. Zhang, B., Al Dhuraibi, Y., Rouvoy, R., Paraiso, F., and Seinturier, L.: CloudGC: recycling idle virtual machines in the cloud. In: Proceedings of 2017 IEEE International Conference on Cloud Engineering, pp. 105–115 (2017)
24. What is CPU steal time? https://www.site24x7.com/learn/linux/cpu-steal-time.html. Accessed 30 Jan 2024
25. PSI – Pressure Stall Information. https://docs.kernel.org/accounting/psi.html. Accessed 30 Jan 2024

Transient Behavior of the Photonic Switch with Duplication of Switching Elements in the All-Optical Network with Heterogeneous Traffic

Konstantin Vytovtov[ID] and Elizaveta Barabanova[(✉)][ID]

V. A. Trapeznikov Institute of Control Sciences of RAS, 65 Profsoyuznaya Street, Moscow, Russia
elizavetaalexb@yandex.ru

Abstract. The transient operating mode of the optical switch with duplication of switching elements in the all-optical network with heterogeneous traffic is investigated. To analyze the performance characteristics of the switch the multi-line queuing system with limited buffer and correlated input flow is investigated. The analytical method of solving the Kolmogorov system of equations based on finding the probability translation matrix is used. The analytical expressions are given for non-stationary probabilities of switch states, and based on them expressions of probability of losses and throughput are presented. In the numerical example, non-stationary and stationary values of state probabilities, throughput and values of transition time for the all-optical switches with two switching elements and buffer size equal to one packet are analyzed. The performance metrics are analyzed in depends on different values of arrival rate.

Keywords: optical switch · throughput · transient mode · correlated input flow · switching element · duplication · translation matrix

1 Introduction

When developing a complex objects that require processing large amounts of information, optical telecommunication systems are used [1]. The perspective switching devices of such a networks are all-optical ones [2]. Their peculiarity is that there is no need to convert optical signals coming from sensors or other technical devices into electrical ones at the input and carry out the reverse conversion at the output. Therefor the performance of such switches becomes significantly higher than their electro-optical analogues. Depending on the method of controlling the optical beam, electro-optical, thermo-optical, acousto-optical

The reported study was funded by Russian Science Foundation, project number 23-29-00795, https://rscf.ru/en/project/23-29-00795/.

and other types of all-optical switching elements are distinguished. The fastest of them are electro-optical systems, which switching speed is about several tens of picoseconds [3]. To build switching systems, basic photonic cells are combined into various types of optical schemes, such as a Clos switch, Banyan network, matrix switch, etc. And to ensure non-blocking of switching process the centralized control algorithms are used. It should be noted that the relatively low speed of the electronic processor, which performs the function of a central control device, negatively affects the overall throughput of an all-optical switch [4]. Therefore, the development of optical switches with decentralized control, which can significantly improve the performance of modern optical all-optical networks is a relevant and important direction.

The other relevant problem of next-generation networks investigation is study of switching systems transient behavior. Transient modes may occur in the result of installation or reboot of the switch. Moreover, if for an electronic switch such modes have a short duration compared to the total packet transmission time, then for an optical switch, due to its high performance, the duration of the transient process is comparable to the packet transmission time, and therefore the study of the performance characteristics of the switch in the transient mode is necessary and more accurately reflects the operation of the device.

This work examines the transient operating mode of optical switches with decentralized control, a feature of which is the presence of several switching elements in one switching unit [5]. The operation of such devices can be described and their performance characteristics can be found using models of multi-line queuing systems [6–8]. It should be noted that the traffic of modern networks is heterogeneous and is more accurately described by a so-called correlated flow. Moreover, if to analyze the stationary characteristics of such models a system of algebraic equations must be solved [6–8]. While to study the transient characteristics the more complex problem of solving differential equations arises. In queuing theory, either numerical methods [9] or the Laplace transform apparatus are used for this, which allows one to find a solution to the system of Kolmogorov differential equations for specific initial conditions [10]. And only a few works present an approach to finding a general solution to the Kolmogorov equations, based on the use of the so-called probability translation matrix [11,12].

This paper examines the transient operating mode of optical switches of all-optical network using the apparatus of queuing theory. The proposed analytical method of the probability translation matrix allows us to obtain analytical expressions for finding the dependencies of the main characteristics of the switch's performance on time in a general form, which is an advantage when solving the problem of synthesis and designing all-optical networks.

The paper consists of the following sections: Sect. 2 provides the statement of the problem, a description of the analytical approach to studying the probability of states of the all-optical switch in transient mode is considered in Sect. 3, the mathematical models of performance metrics are presented in Sect. 4 and Sect. 5 presents results of numerical simulation.

2   The Statement of the Problem

The 8×8 photonic switch is investigated [5]. The circuit includes several switching elements, one of which is the main one, and the rest operate in a hot standby mode. It is assumed that all switching elements serve packets arriving at them in load sharing mode. The problem is to calculate the main performance metrics of a given switch in transient mode, such as probabilities of states, packet loss probability, throughput and transient time.

As modern networks are characterized by correlated information flows the $MAP/M/n/N$ queuing system can be used as the mathematical model for describing functioning of the all-optical switch. According to this model the input flow is described by two nonzero $M \times M$-matrices $\mathbf{D}_0$ and $\mathbf{D}_1$ [13]:

$$
\mathbf{D}_0 = \begin{pmatrix} -\lambda_0 & \lambda_0 p_{0,1}^{(0)} & \cdots & \lambda_0 p_{0,M-1}^{(0)} \\ \lambda_1 p_{1,0}^{(0)} & -\lambda_1 & \cdots & \lambda_1 p_{1,M-1}^{(0)} \\ \cdots & & \cdots & \cdots \\ \lambda_{M-1} p_{M-1,0}^{(0)} & \lambda_{M-1} p_{M-1,1}^{(0)} & \cdots & -\lambda_{M-1} \end{pmatrix} \tag{1}
$$

$$
\mathbf{D}_1 = \begin{pmatrix} \lambda_0 p_{0,0}^{(1)} & \lambda_0 p_{0,1}^{(1)} & \cdots & \lambda_0 p_{0,M-1}^{(1)} \\ \lambda_1 p_{1,0}^{(1)} & \lambda_1 p_{1,1}^{(1)} & \cdots & \lambda_1 p_{1,M-1}^{(1)} \\ \cdots & & \cdots & \cdots \\ \lambda_{M-1} p_{M-1,0}^{(1)} & \lambda_{M-1} p_{M-1,1}^{(1)} & \cdots & \lambda_{M-1} p_{M-1,M-1}^{(1)} \end{pmatrix} \tag{2}
$$

where λ_ν is an exponential distributed input rate, M is the number of states of non-periodic Markov chain $\nu_t, t \geq 0$. Here the probability $p^{(1)}(\nu, \nu')$ is the probability that the process $\nu_t, t \geq 0$ goes to some state ν' and a packet is generated and the $p^{(0)}(\nu, \nu')$ is the probability that the process makes a transition but without generating a packet. The probability that there is a jump from the state ν to the same state is equal to zero therefor $p^{(0)}(\nu, \nu) = 0$.

According to the $MAP/M/n/N$ model the 8×8 switch can function in one of $K = M \times (n+N)$ states (Fig. 1), where n is the number of switching elements and N is the size of the buffer. The states of the graph $S_0, S_1, ..., S_n$ determine the macrostates of the switching elements and the states $S_{n+N-1}, ..., S_{n+N}$ determine the macrostates of the buffer. Transitions between these macrostates are accompanied either by generation of a new packet, or by servicing the old one. Thus, the system is in the S_0 macrostate if there are no packets in it; that is, the first switching element is idle and the buffer is empty. The system is in the S_1 macrostate if the first switching element processes one packet and the next ones $n - 1$ switching elements are idle and the buffer are empty. The system is in the S_n macrostate, if all n switching elements are busy and the buffer is empty. The system is in the S_{n+N-1} macrostate, if all n switching elements are busy and $N - 1$ packet is in the buffer. The system is in the S_{n+N} macrostate if all switching elements are busy, there are N packets in the buffer, and the next incoming packet will be loosed. Each of the above macrostates S_k, where $k = 0, ...n + N$ corresponds to M additional states $S_k^{(0)} ... S_k^{(M-1)}$

of the MAP-flow control process without generating a packet. The transitions between macrostates correspond to transitions of the MAP with packet arrival (M possible transitions) or the service completion with rate $n\mu$ (also M possible transitions between states). A change in the state of the system $S_k^{(0)} \ldots S_k^{(M-1)}$ can occur as a result of one of the $M-1$ state transitions of the control MAP process.

To find the performance metrics of the switch in a transient mode such as probability of loss and throughput the probability translation matrix method is used [11,12].

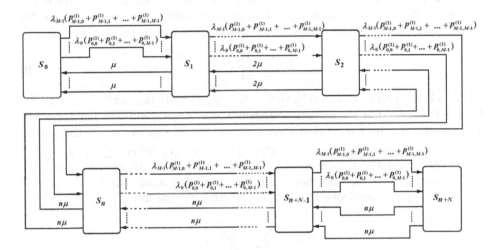

Fig. 1. The graph of $MAP/M/n/N$ system states

3   The Mathematical Model of the Photonic Switch with Duplication of Switching Element

First of all the system of Kolmogorov equations for the $MAP/M/n/N$ system can be write down:

$$\frac{d}{dt}\vec{P} = \mathbf{A}\vec{P} \tag{3}$$

where $\mathbf{A}$ is system coefficient matrix which has the form:

$$\mathbf{A} = \begin{pmatrix} \mathbf{D}_0^T & \mu\mathbf{I} & \ldots & O & O \\ \mathbf{D}_1^T & \mathbf{D}_0^T - \mu\mathbf{I} & \ldots & 2\mu I & O \\ \ldots & \ldots & \ldots & \ldots & \ldots \\ O & \mathbf{D}_1^T & \ldots & \mathbf{D}_0^T - 2\mu I & 2\mu I \\ O & O & \ldots & \mathbf{D}_1^T & \mathbf{D}_0^T + \mathbf{D}_1^T - 2\mu\mathbf{I} \end{pmatrix} \tag{4}$$

In (3) $\boldsymbol{P}(t) = \left(\boldsymbol{P}_0(t), \boldsymbol{P}_1(t), \ldots, \boldsymbol{P}_n(t), \ldots, \boldsymbol{P}_{n+N}(t)\right)^T$ are the probabilities of the serving and buffer device states S_k, where $k = \overline{0, n+N}$ (each S_k has M states: from $S_k^{(0)}$ to $S_k^{(M-1)}$) determined by the dimension of the matrices (1) and (2). Here $\boldsymbol{P}_0(t)$ is the probability that the switch is idle, $\boldsymbol{P}_n(t)$ is the probability that all n switching elements are busy, and $\boldsymbol{P}_{n+N}(t)$ is the probability of packet loss. According to the probability translation matrix method the solution of (3) can be found in the following form:

$$\mathbf{P}(t) = \mathbf{L}(t)\mathbf{P}(t_0) \tag{5}$$

where $\boldsymbol{P}(t)$ is the vector of state probabilities at a given time, $\boldsymbol{P}_0(t)$ is the column vector of initial conditions of state probabilities, $\mathbf{L}(t)$ is the probability translation matrix [11,12], which elements can be found in the following form

$$L_{ij} = \frac{\Delta_{ji}}{\Delta} \zeta_{ij}\gamma_j \tag{6}$$

where γ is the j-th root of the characteristic polynomial $\gamma(b_{K-1}\gamma^{K-1} + \ldots + b_2\gamma^2 + b_1\gamma) = 0$; ζ_{ij} are the elements of the eigen-basis of the coefficient matrix of the system (3); Δ_{ji} are the algebraic complement to the element ζ_{ij} of the matrix (3) for i-vector of initial conditions $\mathbf{P}_{0i} = \{x_1, \ldots, x_{K+1}\}$, where $\sum_{i=1}^{K-1} x_i = 1$; Δ is the determinant of the eigen-basis of the coefficient matrix of the system (3).

For example, the second column of the translation matrix $\mathbf{L}(t)$ can be found under the initial conditions $\mathbf{P}_{02} = \{0, 1, \ldots 0\}$ and has the form:

$$L_{i2} = \sum_{j=0}^{K-1} \frac{\Delta_{2,j}}{\Delta} \zeta_{ij}\gamma_j \tag{7}$$

4   The Mathematical Models of Performance Metrics

The goal the investigation is to obtain analytical expressions for the key transient performance metrics of the all- optical switch, such as a packet loss probability, a throughput, and a transient time.

Probability of Loss. The expression for the translation matrix (6) allows calculating the probability of packet loss at each time instant of the transient mode as the sum of the probabilities of the states $S_{n+N}^{(0)} \ldots S_{n+N}^{(M-1)}$ in this moment in time:

$$P_{loss} = \sum_{i=0}^{M-1} P_{n+N}^{(i)}(t) \tag{8}$$

Using (6) we obtain

$$P_{loss} = \sum_{i=K-M}^{K-1} \sum_{j=0}^{K-1} L_{ij}(t - t_0) P_j(t_0) \tag{9}$$

Transient Time. As it was shown in [11] the transient time is determined by the minimum of $|\alpha|$, where α is the real part of $\gamma = \alpha + \iota\beta$:

$$t_{tr} = k \cdot \frac{1}{|\alpha_{min}|}, \tag{10}$$

where k is the coefficient which determined by the practical requirements for specific queuing systems. In most cases $k = 3 \div 5$.

Throughput. The throughput of the system at a given time in accordance with the definition [11] has the form:

$$A(t) = [1 - P_{loss}(t)] \cdot \lambda = [1 - \sum_{i=0}^{M-1} P_{n+N}^{(i)}(t)] \cdot \lambda \tag{11}$$

Here $\lambda = \theta \mathbf{D}_1 \vec{e}$ is the average rate of the input flow. Taking into account the matrix (6) the result expression for the throughput has the form:

$$A(t) = [1 - \sum_{i=K-M}^{K-1} \sum_{j=0}^{K-1} L_{ij}(t - t_0)P_j(t_0)] \cdot \lambda \tag{12}$$

5   The Numerical Results

In the numerical example the all-optical network with throughput of 1 Gbps and correlated input flows is investigated. It is assumed that the Ethernet input flow with maximum size of 1500 bytes includes three type of traffic is considered (the number of states $M = 3$). The arrival rate of the first type of traffic $\lambda_0 = 9 \cdot 10^3$ pps, the second type of traffic is transferred with the arrival rate $\lambda_1 = 20 \cdot 10^3$ pps, and the arrival rate of the third one $\lambda_2 = 30 \cdot 10^3$ pps. So the values of the flow rates can be set by the matrix $\Lambda = diag\{\lambda_0, \lambda_1, \lambda_2\} = diag\{9 \cdot 10^3, 20 \cdot 10^3, 30 \cdot 10^3\}$. The average rate of the input flow $\lambda = 12.117 \cdot 10^3$ pps. The service rate is constant $\mu = 0.5 \cdot 10^3$ pps. Here we consider the case when the all-optical switch has two switching elements ($n = 2$) and the size of the buffer is equal to one packet ($N = 1$). Thus, the queuing system describing the such a switching system has twelves states.

The graph of the system states is presented in Fig. 2. Here the states "1", "2", and "3" correspond to the free states of all two switching elements. The states "4", "5", and "6" correspond to the states of the first switching element functioning in the cases when it transmits the traffic of the first type with λ_0 (the state "4"), the second type with λ_1 (the state "5") and the third type with λ_2 (the state "6"). The states "7", "8", and "9" correspond to the states when the first and the second switching elements transmit one of three types of traffic. The states "10", "11", and "12" correspond to the states when two switching elements are occupied and one packet is in buffer. The last three states

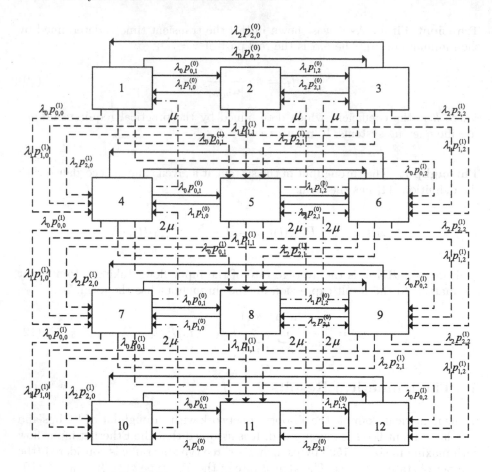

Fig. 2. The graph of $MAP/M/2/1$ system states

correspond to the loss states because the next packet entering the system will be dropped.

The matrixes of transition probabilities values between the system states for the case when the packet arrives in the system and for the case when the packet does not arrive in the system are p_0 and p_1 correspondingly:

$$\mathbf{P}^{(0)} = \begin{pmatrix} 0 & p_{01}^{(0)} & p_{02}^{(0)} \\ p_{10}^{(0)} & 0 & p_{12}^{(0)} \\ p_{20}^{(0)} & p_{21}^{(0)} & 0 \end{pmatrix} = \begin{pmatrix} 0 & 0.1 & 0.1 \\ 0.3 & 0.2 & 0.5 \\ 0.1 & 0.4 & 0.1 \end{pmatrix} \tag{13}$$

$$\mathbf{P}^{(1)} = \begin{pmatrix} p_{00}^{(1)} & p_{01}^{(1)} & p_{02}^{(1)} \\ p_{10}^{(1)} & p_{11}^{(1)} & p_{12}^{(1)} \\ p_{20}^{(1)} & p_{21}^{(1)} & p_{22}^{(1)} \end{pmatrix} = \begin{pmatrix} 0.1 & 0.1 & 0.4 \\ 0.2 & 0.2 & 0.5 \\ 0.3 & 0.3 & 0.3 \end{pmatrix} \tag{14}$$

Thus the matrices $\mathbf{D_0}$ and $\mathbf{D_1}$ describing the input flow have the form according to (1) and (2):

$$\mathbf{D_0} = \begin{pmatrix} -9000 & 900 & 900 \\ 6000 & -20000 & 10000 \\ 3000 & 12000 & -30000 \end{pmatrix} \tag{15}$$

$$\mathbf{D_1} = \begin{pmatrix} 900 & 900 & 3600 \\ 4000 & 4000 & 10000 \\ 9000 & 9000 & 9000 \end{pmatrix} \tag{16}$$

At the first step of the numerical algorithm the matrix $\mathbf{A}$ eigenvalues (3) must be evaluated: $\gamma_0 = 0$, $\gamma_1 = -4.71 \cdot 10^4 + 6295.86\iota$, $\gamma_2 = -4.7 \cdot 10^4 - 6295.86\iota$, $\gamma_3 = -5.694 \cdot 10^3$, $\gamma_4 = -4.2 \cdot 10^4$, $\gamma_5 = -3.9 \cdot 10^4$, $\gamma_6 = -2.29 \cdot 10^4 + 7253.52\iota$, $\gamma_7 = -2.29 \cdot 10^4 - 7253.52\iota$, $\gamma_8 = -1.495 \cdot 10^4$, $\gamma_9 = -1.81 \cdot 10^4$, $\gamma_{10} = -2 \cdot 10^4$, $\gamma_{11} = -2.94 \cdot 10^4$. The analysis of eigenvalues shows that the eigenvalues for this case are either real negative or pairwise complex conjugate with negative real parts. Using (10) the transient time $t_{tr} = 5/\alpha_{min} = 5/5.694 \cdot 10^3 = 0.00088$ s (Fig. 3). Figure 3 shows the dependencies of the state probabilities: p_{loss} is the probability of losses, p_{idle} is the probability that the system is idle, p_1 is the probability that one switching element is busy and p_2 is the probability that all switching elements are busy. According to the Fig. 3 the stationary probabilities of the states have the following values: $\pi_{loss} = 0.18$, $\pi_{idle} = 0.23$, $\pi_1 = 0.33$, $\pi_2 = 0.26$. The sum of the probabilities of all states is equal to one.

In the Fig. 4 the dependencies of the state probabilities for the values of input rates specified by the matrix $\Lambda = diag\{8 \cdot 10^3, 10 \cdot 10^3, 10 \cdot 10^3\}$ pps and $\mu = 0.5 \cdot 10^3$ pps are presented. The average arrive rate decreased compared to the first case $\lambda = 6.663 \cdot 10^3$ pps. The values of eigenvalues of the $\mathbf{A}$ (3) are the following: $\gamma_0 = 0$, $\gamma_1 = -4.91 \cdot 10^3$, $\gamma_2 = -2.56 \cdot 10^4 + 2286.22\iota$, $\gamma_3 = -2.56 \cdot 10^4 - 2286.22\iota$, $\gamma_4 = -1.771 \cdot 10^4 + 6574.85\iota$, $\gamma_5 = -1.771 \cdot 10^4 - 6574.85\iota$, $\gamma_6 = -2.115 \cdot 10^4$, $\gamma_7 = -1.93 \cdot 10^4$, $\gamma_8 = -1.149 \cdot 10^4$, $\gamma_9 = -1.31 \cdot 10^4 + 1337.9\iota$, $\gamma_{10} = 1.31 \cdot 10^4 - 1337.9\iota$, $\gamma_{11} = -1.536 \cdot 10^4$. Using (10) it can be shown that the transient time increased $t_{tr} = \frac{5}{|\alpha_{min}|} = 5/4.91 \cdot 10^3 = 0.001$ s. The decreasing of the average arrive rate influenced to the values of stationary state probabilities: $\pi_{loss} = 0.08$, $\pi_{idle} = 0.41$, $\pi_l = 0.34$, $\pi_2 = 0.17$. As it can be seen the probability of loss decreased.

In Fig. 5 the dependencies of the throughput of the switching system for two cases are presented. In the left figure the dependence of the throughput for the first case is shown ($\lambda = 12.117 \cdot 10^3$ pps) and in the right figure the dependence of the throughput for the second case is presented ($\lambda = 6.663 \cdot 10^3$ pps).

The stationary throughput in the first and second cases are 13740 pps and 6920 pps correspondingly and can be calculated using the following expression:

$$A(t \to \infty) = (1 - \pi_{loss}) \cdot \lambda \tag{17}$$

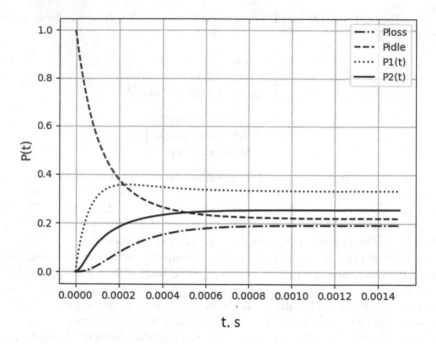

Fig. 3. The dependencies of the state probabilities

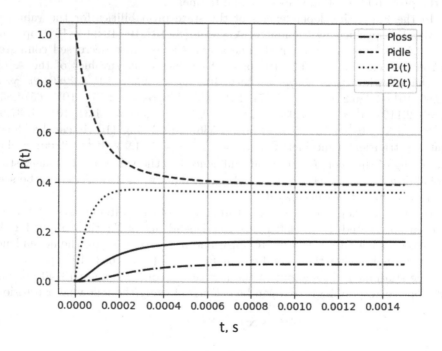

Fig. 4. The dependencies of the state probabilities

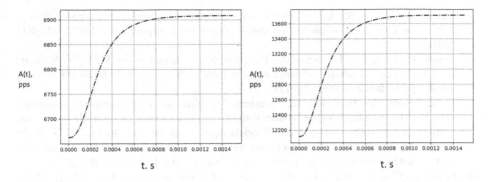

Fig. 5. The dependencies of the throughput

6 Conclusion

The main goal of this work is to study the transient mode of the all- optical switch with duplicating switching elements for case of correlated input flows that imitated the traffic of a real optical network. The queuing system $MAP/M/n/N$ is used for described the functioning of this switch. The method of probability translation matrix is applied for finding analytical expressions of basic performance metrics of all-optical switches with duplication switching elements such a probability of loss, throughput and transient time. The numerical results of performance metrics calculation for the system with two duplicated switching elements using $MAP/M/2/1$ model are presented. It is shown how a decrease of the input flow intensity affects the probabilities of system states and the throughput of the all-optical switch. The results of this study allow to analyze the characteristics of the all-optical switch in various operating modes and can be used in the design of next-generation network all-optical switches.

References

1. Nandi, D., Nandi, S., Sarkar, A., Sarkar, C.K.: Optical Switching: Device Technology and Applications in Networks. Wiley, Hoboken (2022)
2. Yong, L., Qibo, F., Lishuang, L., Qingrui, Y., Yueqiang, L.: Application of optical switch in precision measurement system based on multi-collimated beams. Measurement **61**, 216–220 (2015)
3. Bawab, E.: Optical Switching. Springer Science and Business Inc., New York (2006). https://doi.org/10.1007/0-387-29159-8
4. Xue, X., Calabretta, N.: Nanosecond optical switching and control system for data center networks. Nat. Commun. **13**, 2257 (2022)
5. Barabanova, E., Vytovtov, K., Podlazov, V.: Model and algorithm of next generation optical switching systems based on 8 × 8 elements. In: Vishnevskiy, V.M., Samouylov, K.E., Kozyrev, D.V. (eds.) DCCN 2019. LNCS, vol. 11965, pp. 58–70. Springer, Cham (2019). https://doi.org/10.1007/978-3-030-36614-8_5

6. Kumar, S., Sunny, S.: $M/M/c/N$ queuing systems with encouraged arrivals, reneging, retention and Feedback customers. Yugoslav J. Oper. Res. **28**(3), 333–344 (2018)
7. Smith, P., Firag, A., Dmochowski, P., Shafi, M.: Analysis of the $M/M/N/N$ queue with two types of arrival process: applications to future mobile radio systems. J. Appl. Math. **2012**, 123808 (2012)
8. Bouchentouf, A., Medjahri, L., Boualem, M., Kumar, A.: Mathematical analysis of a Markovian multi-server feedback queue with a variant of multiple vacations, balking and reneging. Discr. Cont. Models Appl. Comput. Sci. **30**(1), 21–38 (2022)
9. Singla, N., Garg, P.C.: Transient and numerical solution of a feedback queueing system with correlated departures. Am. J. Numer. Anal. **2**(1), 20–28 (2014)
10. Sah, S.S., Ghimire, R.P.: Transient analysis of queueing model. J. Inst. Eng. **11**(1), 165–171 (2015)
11. Vishnevsky, V., Vytovtov, K., Barabanova, E., Semenova, O.: Transient behavior of the $MAP/M/1/N$ queuing system. Mathematics **9**, 2559 (2021)
12. Vyshnevsky, V.M., Vytovtov, K.A., Barabanova, E.A., Semenova, O.V.: Analysis of an $MAP/M/1/N$ queue with periodic and non-periodic piecewise constant input rate. Mathematics **10**(10), 1684 (2022)
13. Dudin, A.N., Klimenok, V.I., Vishnevsky, V.M.: Methods to study queuing systems with correlated arrivals. In: The Theory of Queuing Systems with Correlated Flows, pp. 63–146. Springer, Cham (2020). https://doi.org/10.1007/978-3-030-32072-0_2

FPGA Implementation of a Decoder with Low-Density Parity Checks Based on the Minimum Sum Algorithm for 5G Networks

Dmitry Aminev[1], Rostislav Danilov[1], and Dmitry Kozyrev[2,3]([⊠]) [iD]

[1] MIREA – Russian Technological University, 78 Vernadsky Avenue, Moscow 119454, Russia
aminev.d.a@ya.ru

[2] V. A. Trapeznikov Institute of Control Sciences of Russian Academy of Sciences, 65 Profsoyuznaya Street, Moscow 117997, Russia

[3] Peoples' Friendship University of Russia (RUDN University), 6 Miklukho-Maklaya Street, Moscow 117198, Russian Federation
kozyrev-dv@rudn.ru

Abstract. The scope of application of codes with low density parity checks and their role in the 5th generation mobile communication networks are considered. The mathematical apparatus of coding with low density parity checks is disclosed. A generalized and detailed decoder architecture optimized for solutions for 5G mobile networks is proposed. Fragments of source code in field-programmable gate array (FPGA) Verilog programming language are presented. In the Xilinx Vivado development environment, using developed test programs, modeling of both some individual project modules and the decoder macromodule was carried out. The results of compiling the decoder project with an analysis of the involved resources of the selected FPGA are presented.

Keywords: low density parity code · error correction code · Modelsim · Xilinx · Verilog · FPGA · 5G mobile network

1 Introduction

5G New Radio (NR), defined by the 3GPP Project, is a standard designed to meet the requirements of a variety of application scenarios, including enhanced mobile broadband (eMBB), ultra-reliable low-latency communications (URLLC) and Massive Machine-Type Communications (mMTC) [1,2]. Unlike the turbo code used in 4G (LTE) and LTE-Enhanced (LTE-A), low-density parity check code (LDPC) is used in 5G NR as the basis of data channel encoding [3], which

This paper has been supported by the RUDN University Strategic Academic Leadership Program and funded by the Russian Science Foundation according to the research project No. 22-49-02023.

V. M. Vishnevskiy et al. (Eds.): DCCN 2023, CCIS 2129, pp. 57–76, 2024.
https://doi.org/10.1007/978-3-031-61835-2_5

helps improve data rates , delays, compatibility and other performance indicators [4–6].

Low-density parity-check code (LDPC code), used in data transmission systems, is a special case of block linear parity-check code [7]. Its feature is the low density of significant elements of the check matrix, due to which the relative simplicity of implementing coding tools is achieved [8].

LDPC codes were first described by Robert Gallager in 1963 in his dissertation [9], but due to the lack of high-performance computing tools, they did not find practical application. It was only in the 1990s that methods for constructing LDPC codes with reduced complexity were proposed by D. J. C. MacKay and R. M. Neal [10]. In these linear block codes, the check bits are located at the end of the information message. An example of the use of LDPC codes in a data transmission path over a noisy channel is presented in Fig. 1.

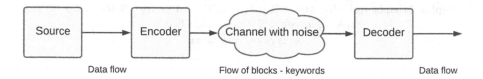

Fig. 1. Data transmission path over a noisy channel

At the moment, LDPC codes are utilized in Wi-Fi standards, LTE and 5G mobile radio access network standards [11]. LDPC codes are used in applications such as digital satellite broadcasting (DVB-S2), wireless local area networks (IEEE 802.11n) and wireless metropolitan area networks (IEEE 802.16e) [12].

The LDPC code is a linear block code that is defined by an $N \times M$ sparse parity check matrix H. N denotes the number of bits in a codeword (or a block), and M is the number of parity checks. The matrix defining the LDPC code must be sparse, which implies a low density of units and a low density of significant elements, and a relatively large dimension [13].

The LDPC decoder can be based on the bit-flipping algorithm, sum-product algorithm, and min-sum algorithm [14].

When implemented on FPGAs in network systems, it is advisable to use the minimum sum algorithm, since the algorithm obtained by simplifying the calculation of messages from check nodes to bit nodes greatly simplifies the implementation of decoding tools [15].

2   Mathematical Apparatus for Performing Low-Density Parity-Check Encoding

2.1   Encoding Principle

Encoding is multiplication of an information message vector of length K by a generating matrix G:

$$a_{1 \times N} = u_{1 \times K} \otimes G_{K \times N}, \tag{1}$$

where $\otimes$ denotes modular multiplication. The generating matrix consists of two concatenated (connected) parts:

$$G_{K \times N} \left[I_{K \times K} - P_{K \times (N-K)} \right], \tag{2}$$

where P — even part, I — identity matrix.

The rows of the generating matrix must be linearly independent.

The generating matrix is directly related to another critical matrix used during the decoding procedure: the parity check matrix. The parity check matrix has (N-K) rows and N columns, where N corresponds to the required codeword length and K corresponds to the message length:

$$H_{(N-K) \times N} \left[P^T_{(N-K) \times K} - I_{(N-K) \times (N-K)} \right]. \tag{3}$$

That is, there are two types of nodes: so-called variable nodes, the number of which corresponds to the number of columns K, and check nodes, corresponding to the number of rows $(N - K)$. The nodes are connected to each other, and the connection is determined by the position of the units in the matrix H.

In order to consider the decoding procedure successful, it is necessary that certain values – usually zeros – are generated at all check nodes.

$$S = H \times x = 0. \tag{4}$$

Here matrix S is a parity check syndrome, and x is a vector of the encoded message. Parity check matrices for LDPC codes must be sparse, i.e. they must contain significantly more zeros than non-zero elements.

2.2 Decoding Principle

Representation of noisy values obtained from a communication channel in the form of so-called soft demodulation values (soft decisions) is performed in the form of log likelihood ratio (LLR):

$$r = \ln \left(\frac{\mathbb{P}(x = 0)}{\mathbb{P}(x = 1)} \right) = \ln \left(\frac{1 - p}{p} \right),$$

where p denotes probability and x is some event.

Consider 2 types of messages: Variable-to-Check (V2C) message — a message from check nodes to variable nodes, and vice versa — Check-to-Variable (C2V) message.

Decoding order is as follows:

1. Initialization
   The starting point for our algorithm is the matrix of LLR values, repeating the structure of the matrix H. Let us select an analytical description:

$$M_{M \times N} = (r_{N \times 1} \cdot 1_{1 \times N})^T \circ H_{M \times N},$$

where 1 is an array of ones and ∘ denotes the Hadamard product (element-wise multiplication). In practice, we can do without the identity matrix: we replace the bracket with iterative Hadamard multiplication of the LLR vector with the columns of the parity check matrix (an additional loop will be needed). If the matrices are large enough, this approach may be more memory efficient.

2. V2C message

Then comes the so-called horizontal step: the algorithm requires processing of the V2C message in the domain of probability. To move from LLR to probabilities, we use the relationship between the hyperbolic tangent and the natural logarithm.

$$\tanh\left(\frac{1}{2}\ln\left(\frac{1-p}{p}\right)\right) = 1 - 2p. \tag{5}$$

Strictly speaking, the procedure for transmitting a V2C message is the multiplication of non-zero probabilities in each line:

$$E_{j,i} = \log\left(\frac{1 + \prod_{i' \in B_{j,i' \neq i}} \tanh\left(\frac{M_{j,i'}}{2}\right)}{1 - \prod_{i' \in B_{j,i' \neq i}} \tanh\left(\frac{M_{j,i'}}{2}\right)}\right) = \log\left(\frac{1 + \prod_{i' \in B_{j,i' \neq i}} M_{j,i'}}{1 + \prod_{i' \in B_{j,i' \neq i}} M_{j,i'}}\right),$$

where j is the number of a particular row, i is the number of a particular column, B_j is the set of non-zero values in the j-th row, and the expression $i' \neq i$ means that we exclude the i-th variable node from consideration. Thus, at this stage we need:

  – select an element in the matrix M, transferred to the probabilistic domain;
  – if its position corresponds to the position of a non-zero element in the matrix H, multiply all non-zero probabilities in the row of this element;
  – exclude this element from multiplication (before or after the previous paragraph);
  – repeat all previous steps for each element of the matrix.

3. Checking the Decoding Stop Criteria

At the end of the first iteration, we need to update our prior probabilities - make them posterior:

$$l_i = r_i + \sum_{j \in A_i} E_{j,i},$$

where A_i is the set of elements corresponding to the non-zero elements of the parity check matrix in the i-th column.

We apply them to the bits through the reverse Non-return-to-zero (NRZ) code:

$$Z_i = \begin{cases} 0, \text{if } l_i \geq 0, \\ 1, \text{if } l_i < 0. \end{cases}$$

Further we calculate the syndrome using formula (4). If the vector is zero, we stop decoding. If not, then move on to the next step.

4. C2V message
Calculate

$$M_{j,i} = \sum_{j' \in A_{i,j'} \neq j} E_{j',i+r_i}.$$

Further we proceed to calculating the matrix E. And so on until step 3 is completed (or the number of available iterations ends).

2.3 Minimum Sum Algorithm for LDPC Decoder

When decoding LDPC codes, many algorithms are used, one of the simplest and at the same time effective is the Min-Sum algorithm. Decoding can be divided into 4 stages, shown in Fig. 2:

1) Initialization
2) Operation on strings
3) Operation on columns
4) Final conclusion

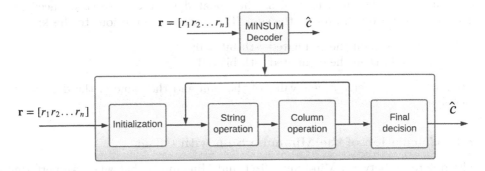

Fig. 2. Model of an LDPC decoder based on the Min-Sum algorithm

During initialization, an auxiliary matrix L is created in which the elements that were zeros in the check matrix H remain zeros, but all ones are replaced with potential non-zero x values.

$$H = \begin{pmatrix} 1 & 1 & 1 & 0 & 1 & 0 & 0 \\ 0 & 1 & 1 & 1 & 0 & 1 & 0 \\ 1 & 1 & 0 & 1 & 0 & 0 & 1 \\ 1 & 0 & 1 & 0 & 1 & 1 & 1 \end{pmatrix}, \quad L = \begin{pmatrix} x & x & x & 0 & x & 0 & 0 \\ 0 & x & x & x & 0 & x & 0 \\ x & x & 0 & x & 0 & 0 & x \\ x & 0 & x & 0 & x & x & x \end{pmatrix}.$$

The decoder receives a message and each value is written to the corresponding column, in the rows where a "potentially non-null value" is present.

$$r = [r_1\, r_2\, r_3\, r_4\, r_5\, r_6\, r_7], \quad L = \begin{pmatrix} r_1 & r_2 & r_3 & 0 & r_5 & 0 & 0 \\ 0 & r_2 & r_3 & r_4 & 0 & r_6 & 0 \\ r_1 & r_2 & 0 & r_4 & 0 & 0 & r_7 \\ r_1 & 0 & r_3 & 0 & r_5 & r_6 & r_7 \end{pmatrix}$$

At this point, the initialization step is considered complete.

The operation on a row can be divided into two stages: finding two minimum values and determining the sign. There are 2 approaches to determining the two minimum values from a series of numbers: "sorting" and "tree", which is better suited when implemented on an FPGA. According to the theory of S.S. Kislitsyn [15], let $V_t(n)$ be the minimum number of comparisons required to determine the t-th largest of n elements, for $1 < t < n$, and let $W_t(n)$ be the minimum number necessary to determine the smallest, the second largest, ..., and the t-th largest value in total. We have:

$$V_t(n) = W_1(n), \; V_t(n) \leq W_t(n), \; W_n(n) = W_{n-1}(n) = S(n).$$

Thus we got the formula:

$$V_2(n) = W_2(n) = n - 2 + [\lg n], \text{ for } n \geq 2.$$

The operation on columns also consists of two stages: finding the sums and defining a new value in the auxiliary table L. After the operation on the columns, we will receive the sum, as well as the updated matrix L. Next we need to determine how we will process the sum. There are two conditions to check:

– if $Sum_j > 0$, then the estimated j-th bit $= 0$;
– if $Sum_j < 0$, then the estimated j-th bit $= 1$.

After which we receive a new value of the sum and the value of the decrypted message.

2.4   Calculation of the Minimum Sum with Offset

The difference between Min-Sum Offset and Min-sum is that when we perform an operation on rows, filling all non-zero places in the row with the first minimum, and replacing the first minimum with the second minimum, we produce an offset by a certain amount β, which is in range $[0, 1]$. After which a check operation occurs, if the transmitted value becomes negative, then it is replaced by zero. This modification of the algorithm is multi-level decoding and allows us to increase the accuracy of calculations.

One of the main problems with decoding LDPC codes is that it takes a lot of time to fully process the rows, and the resources that are needed for operations on the columns are idle. In connection with this problem, an improvement for decoding was invented — multi-level decoding.

Multi-level decoding can be divided into N steps:

1) Initialization of matrix L with grouping of parity check matrix layers,

$$H = \begin{pmatrix} 1\,1\,1\,0\,1\,0\,0 \\ 0\,1\,1\,1\,0\,1\,0 \\ 1\,1\,0\,1\,0\,0\,1 \\ 1\,0\,1\,0\,1\,1\,1 \end{pmatrix}, \quad H = \begin{pmatrix} H1 \\ H2 \end{pmatrix} = \begin{pmatrix} 1\,1\,1\,0\,1\,0\,0 \\ 0\,0\,0\,1\,0\,1\,1 \\ 1\,1\,0\,1\,0\,0\,1 \\ 0\,0\,1\,0\,1\,1\,0 \end{pmatrix} \begin{matrix} \\ \text{Layer 1} \\ \text{Layer 2} \\ \\ \end{matrix}$$

In this step, the main rule should hold that the weight of the columns of each layer must be equal to one. Of course, one may not follow this rule, but then this will lead to more complicated calculations. After this, we fill the matrix in the same way as we did within the Min-Sum algorithm, but in the first iteration we will consider only the first layer.

$$r = [0.2\ -0.3\ 1.2\ -0.5\ 0.8\ 0.6\ -1.1], \quad L = \begin{pmatrix} 0.2 & -0.3 & 1.2 & 0 & 0.8 & 0 & 0 \\ 0 & 0 & 0 & -0.5 & 0 & 0.6 & -1.1 \\ - & - & 0 & - & 0 & 0 & - \\ 0 & 0 & - & 0 & - & - & 0 \end{pmatrix}.$$

Further, the operations that were described in the Min-Sum algorithm are performed.

After performing operations on the first layer, we move on to the second layer, assuming that the first layer has been completely processed. Operations are performed to find two minimum values and replace values in the matrix L. After which a new sum vector is formed by adding the previous vector, which was obtained after the operation on the first layer, and the values of the corresponding columns of the second layer. When starting a new iteration of processing the matrix L, it is necessary to subtract the first layer from the sum vector obtained in the previous iteration. And refill the first layer. After which all operations on the first layer are repeated, just as in the previous iteration.

The operations on the second layer are identical, i.e. first, a subtraction operation occurs, after which a new sum vector and a new second layer are formed. And the standard operation of finding two minimums and replacing the values in the rows with them. Determining the final solution occurs similarly to the operation described in the Min-Sum algorithm. There are two conditions to check:

- if $Sum_j > 0$, then the estimated j-th bit = 0;
- if $Sum_j < 0$, then the estimated j-th bit = 1.

Thus, in accordance with the minimum sum algorithm for 5G LDPC, the transformation of matrix H is completely performed for one of its rows, that is, the "height" of layers (Layer) in the decoder is equal to 1. In the example of the original matrix H of size 4×7, Layer 1 is provided for processing rows 1 and 2, and Layer 2 — for lines 3 and 4, respectively. This solution, focused on processing one line, avoids the use of FPGA buffer memory with accompanying resource-intensive synchronization and address generation circuits, which generally speeds up the execution of the algorithm (does not introduce buffering time delays) and saves FPGA resources.

2.5   Architecture of LDPC Decoder Using Min Sum Algorithm

The architecture of the LDPC decoder [10], shown in Fig. 3 consists of three main units: Minimum Value Computation Unit (MVCU); Sign Computation

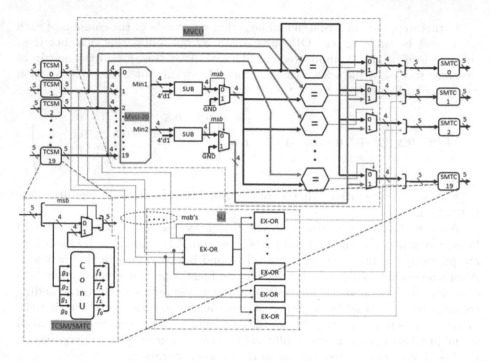

Fig. 3. LDPC decoder architecture

Blocks (SUB); Two's Complement to Sign Magnitude (TCSM) units and Sign Magnitude to Two's Complement (SMTC) units.

The 20-input MVCU consists of an array of MVUs with two inputs (Fig. 3). MVU compares two 5-bit inputs and outputs the minimum value and the index.

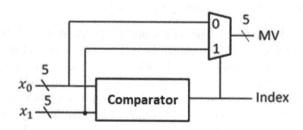

Fig. 4. 2-Input MVU Architecture

The *Index* value is calculated by comparing two inputs using a comparator, and the *MV* is calculated based on the index value using a 2:1 multiplexer:

$$Index = \begin{cases} 0, & \text{if } x_0 < x_1, \\ 1, & \text{otherwise.} \end{cases} \quad ; \quad MV = \begin{cases} x_0, & \text{if } Index = 0, \\ x_1, & \text{otherwise.} \end{cases}$$

Since the original architecture requires a 20-input MVU (MVU-20), it can be constructed from nineteen 2-input MVUs and an adder, as shown in Fig. 5. Ten 2-input MVUs perform the primary input and calculate the minimum value (step A), the rest nine MVUs with two inputs take intermediate minimum values and calculate min_1 (step B). The 5-bit adder adds $Alpha(\alpha)$ to min_1 for calculation of min_2 (step C).

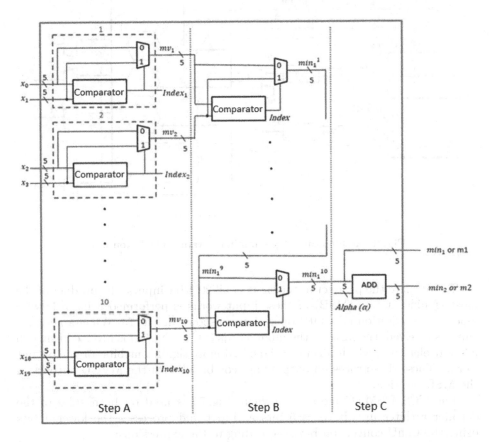

Fig. 5. 20-Input Minimum Value Unit Architecture

The 4-input MVU, highlighted in Fig. 4, consists of three 2-input MVUs and an adder. Two MVUs with two inputs are used to calculate the minimum values mv_1 and mv_2 from the given inputs, and the third MVU is used to calculate min_1. The second minimum value min_2 is calculated by adding a small value of $Alpha(\alpha)$ to the first minimum min_1.

MVU-20 as part of the MVCU, the architecture of which is shown in Fig. 6, calculates the first min_1 and the second min_2. Then the SUB within the MVCU subtracts the offset value 4'd1 from min_1 and min_2, checking the most significant bit (MSB) to determine the sign of the subtracted value. If the subtracted value is negative, then the output of 2:1 Max is zero, otherwise it is the actual subtracted value. Now the entire MVCU output is replaced with min_1, except for the index min_1; this value is replaced by min_2. This is done using 20 equalizers and 2:1 multiplexers.

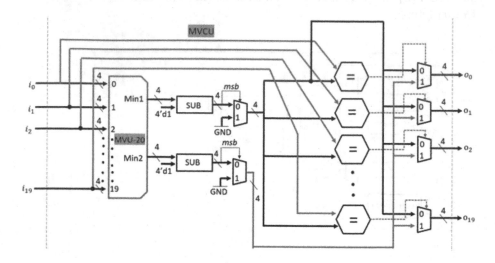

Fig. 6. Architecture of the minimum value calculation unit

To calculate the product of the signs of all 20 4-bit inputs, the product of the most significant bits (MSB) of these input words is performed using a bitwise modulo two addition (EX-OR) operation, as shown in Fig. 7a. When a negative value is received, the MSB of the input words (SU in Fig. 7a) is inverted using 20 EX-OR elements, which form the signed 20 most significant bits for the output words. These sign-processed outputs are combined with the MVCU outputs in the MSB position.

The TCSM/SMTC module shown in Fig. 7b is used on both sides of the decoder architecture. It takes 5 bits as input and processes the lower 4 bits using the ConU conversion block according to the expressions:

$$f_0 = g_0; \quad f_1 = g_1 \oplus g_0; \quad f_2 = (\bar{g}_3 \cdot \bar{g}_2)|(\bar{g}_2 \cdot g_1)|(\bar{g}_2 \cdot \bar{g}_0)|(g_2 \cdot \bar{g}_1 \cdot \bar{g}_0); \quad f_3 = \bar{g}_3|(\bar{g}_2 \cdot \bar{g}_1 \cdot \bar{g}_0).$$

The output of this module is either the actual value or a converted value based on the MSB value, as shown in Fig. 7b.

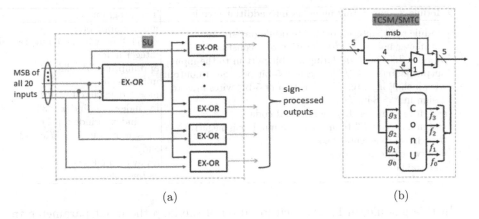

Fig. 7. Architecture of sign calculation units (a) and TCSM/SMTC module (b)

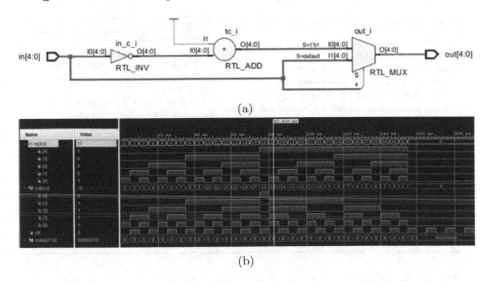

Fig. 8. Synthesized circuit using Tcsm_smtc.v and timing diagrams

2.6 LDPC Decoder Simulation

First, the TCSM/SMTC units (Fig. 7a) and the units for finding the smaller number of the two are simulated, then the entire decoder architecture is simulated.

Module for converting numbers into additional code	Test program

module tcsm_smtc(in, out); //module declaration
parameter nob = 4; //declaration of a constant
input [nob:0] in; //declaring a 5-bit port in to the input
output [nob:0] out; //declaring a 5-bit port out to output
wire [nob:0] in_c, tc; //declaration of 5-bit wires in_c, tc
assign in_c = ~in; // reverse code
assign tc = in_c + 1; // additional code
assign out = in[4]? tc:in; // if the number is negative,
then in complementary code,
otherwise in direct code
endmodule
	module app_tb();
	reg [4:0] in; wire [4:0] out; reg clk;
	integer index;
	app uup(in, out, clk);
	initial begin
	in = 5'd00000; clk = 0;
	for(index = 0; index <32;
	   index = index + 1)
	  begin in = in + 1; #5; end
	end
	always #5 clk=~clk;
	endmodule

In this module, in Fig. 8, each iteration of the loop the input parameter **in** is increased by 1. For each new parameter **in**, additional code is defined.

Module for finding the smaller number of two	Test program

module min_sum_tree_2(V1,V2,min1,min2,ip);
  //module declaration
parameter nob = 4; // input width
input [nob:0] V1, V2; //declaring two
  5-bit ports V1 and V2 as input
output [nob:0] min1, min2; // declaration of two
  5-bit ports min1 and min2 for output
output ip;
reg ip; // register declaration (default value is x)
assign min1 = ip? V2:V1; // if ip=1, then
  min1 = V2, otherwise min1=V1
assign min2 = min1+ 1'b1; // add one to min1 in
  binary code
always @(V1, V2) begin
if (V1 >= V2) ip = 1; else ip = 0; end
endmodule
	module test3_tb();
	reg [4:0] V1, V2; reg clk;
	wire [4:0] min1, min2; wire ip; integer index;
	test3 uup(V1, V2, min1, min2, ip, clk);
	initial begin
	V1 = 5'b00000; V2 = 5'b00000; clk = 0;
	for(index = 0; index <9; index = index + 1)
	  begin
	  if(index % 2 == 0)
	    begin V1 = V1 + 1; V2 = V1 - 1; end
	  else
	    begin V2 = V2 + 1; V1 = V2 - 1; end
	  #5; V1 = V1 + 1; V2 = V2 + 1;
	  end
	end
	always #5 clk=~clk;
	endmodule

The module for finding the smaller number of two and the timing diagrams of its operation are presented in Fig. 9. In the diagrams, the input values **V1** and **V2** alternate each iteration of the loop. The output values output **min1** and **min2** respectively.

The source code of the upper-level LDPC decoder module, corresponding to the circuit in Fig. 2, is presented below in Table 1, and the timing diagrams of its operation are in Fig. 10.

Source code of the test program for the upper-level LDPC decoder module is presented below in Table 2.

Table 1. Source code of the upper-level LDPC decoder module

```
module MSAU(i0, i1,..., i19, o0, o1, ...,o19,min1_index, clk);
parameter nob = 4; // input width
input [nob:0] i0, i1, i2, i3, i4, i5, i6, i7, i8, i9, i10, i11, i12, i13, i14, i15,i16,i17,i18, i19; input clk;
output [nob:0] o0, o1, o2, o3, o4, o5, o6, o7, o8, o9, o10, o11, o12, o13, o14, o15,o16,o17,o18,o19;
output [4:0] min1_index; // current value of the minimum index 1
wire [nob:0] ow0, ow1, ...,ow19; wire [nob:0] iw0, iw1, ...,iw19; wire [nob:0] min1_w, min2_w;
wire [4:0] min1_index_w; // current value of the minimum index 1
wire product_bit;
reg product_bit_0, product_bit_1, ... product_bit_18;
wire [nob:0] min1_w_intermediate, min2_w_intermediate;
wire [nob:0] m1, m2; wire [nob:0] iw0_w, iw1_w,... ,iw18_w;

tcsm_smtc t0(i0, ow0); tcsm_smtc t1(i1, ow1); ... tcsm_smtc t19(i19, ow19); // the value of the
    19th input must be set to the largest value, otherwise an error will occur

min_sum_tree_20 mVG_19(ow0[nob:0], ow1[nob:0], ow2[nob:0], ow3[nob:0], ow4[nob:0], ow5[nob:0],
    ow6[nob:0 ], ow7[nob:0], ow8[nob:0], ow9[nob:0], ow10[nob:0], ow11[nob:0], ow12[nob:0],
    ow13[nob:0], ow14[nob:0], ow15[nob:0],ow16[nob:0],ow17[nob:0],ow18[nob:0],ow19[nob:0],
    min1_w, min2_w, min1_index_w);

assign product_bit = ow0[nob] ^ow1[nob] ^ow2[nob] ^ow3[nob] ^ow4[nob] ^ow5[nob] ^ow6[nob]
    ^ow7[nob] ^ow8[nob] ^ow9[ nob] ^ow10[nob] ^ow11[nob] ^ow12[nob] ^ow13[nob] ^
    ow14[nob] ^ow15[nob] ^ow16[nob] ^ow17[nob] ^ow18[nob];// calculating the sign on one line

always @(negedge clk)
begin
product_bit_0 = product_bit ^ow0[nob]; product_bit_1 = product_bit ^ow1[nob];...;
product_bit_18 = product_bit ^ow18[nob];// sign change for all input values
end

assign min1_w_intermediate = min1_w - 1; //
assign min2_w_intermediate = min2_w - 1; // shift two values by b

assign m1 = min1_w_intermediate[nob]? 5'd0 : min1_w_intermediate;
assign m2 = min2_w_intermediate[nob]? 5'd0 : min2_w_intermediate;// if the minimum with
    a shift is negative (first sign = 1), then replace the minimum with 0, otherwise replace
    it with the minimum with a shift

assign iw0 = (m1 == i0)? m2:m1; assign iw1 = (m1 == i1)? m2:m1;
assign iw2 = (m1 == i2)? m2:m1; assign iw3 = (m1 == i3)? m2:m1; ...;
assign iw18 = (m1 == i18)? m2:m1;// change all values to min1, and min1 to min2

assign iw0_w = {product_bit_0, iw0}; assign iw1_w = {product_bit_1, iw1}; ...;
assign iw18_w = {product_bit_18, iw18}; //change the sign of all values

tcsm_smtc s0 (iw0_w, o0); tcsm_smtc s1 (iw1_w, o1); ... ; tcsm_smtc s18(iw18_w, o18);

assign min1_index = min1_index_w;
endmodule
```

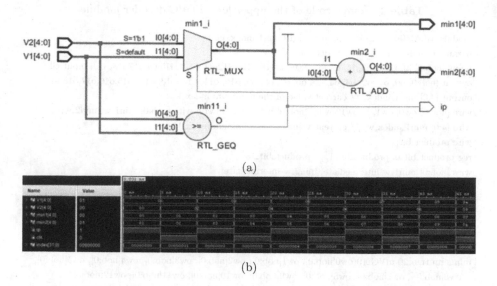

(a)

(b)

Fig. 9. Synthesized digital circuit of the module for finding the smaller number of two (a) and timing diagrams of its operation (b)

Table 2. Source code of the test program for the upper-level LDPC decoder module

```
module MSAU_tb();
reg [4:0] i0, i1, i2, i3, i4, i5, i6, i7, i8, i9, i10, i11, i12, i13, i14, i15,i16,i17,i18, i19;
wire [4:0] o0, o1, o2, o3, o4, o5, o6, o7, o8, o9, o10, o11, o12, o13, o14, o15,o16,o17,o18,o19;
wire [4:0] min1_index;
reg clk;
integer count;
MSAU uut(i0, i1,..., i19, o0, o1, ...,o19,min1_index, clk);
initial begin
i0 = 5'd0; i1 = 5'd0; i2 = 5'd0; i3 = 5'd0; i4 = 5'd0; i5 = 5'd0; i6 = 5'd0; i7 = 5'd0; i8 = 5'd0;
  i9 = 5'd0; i10 = 5'd0; i11 = 5'd0; i12 = 5'd0; i13 = 5'd0; i14 = 5'd0; i15 = 5'd0; i16 = 5'd0;
  i17 = 5'd0; i18 = 5'd0; i19 = 5'd0;
clk = 0;
for(count = 0; count <5; count = count + 1)
  begin
  i0 = 2 + unsigned(random) % (15 - 2); i1 = 2 + unsigned(random) % (15 - 2);...;
  i19 = 2 + unsigned(random) % (15 - 2); // generation of pseudo-random numbers in the
    range from 2 to 15
  #10;
  end
end
always #5 clk=~clk;
endmodule
```

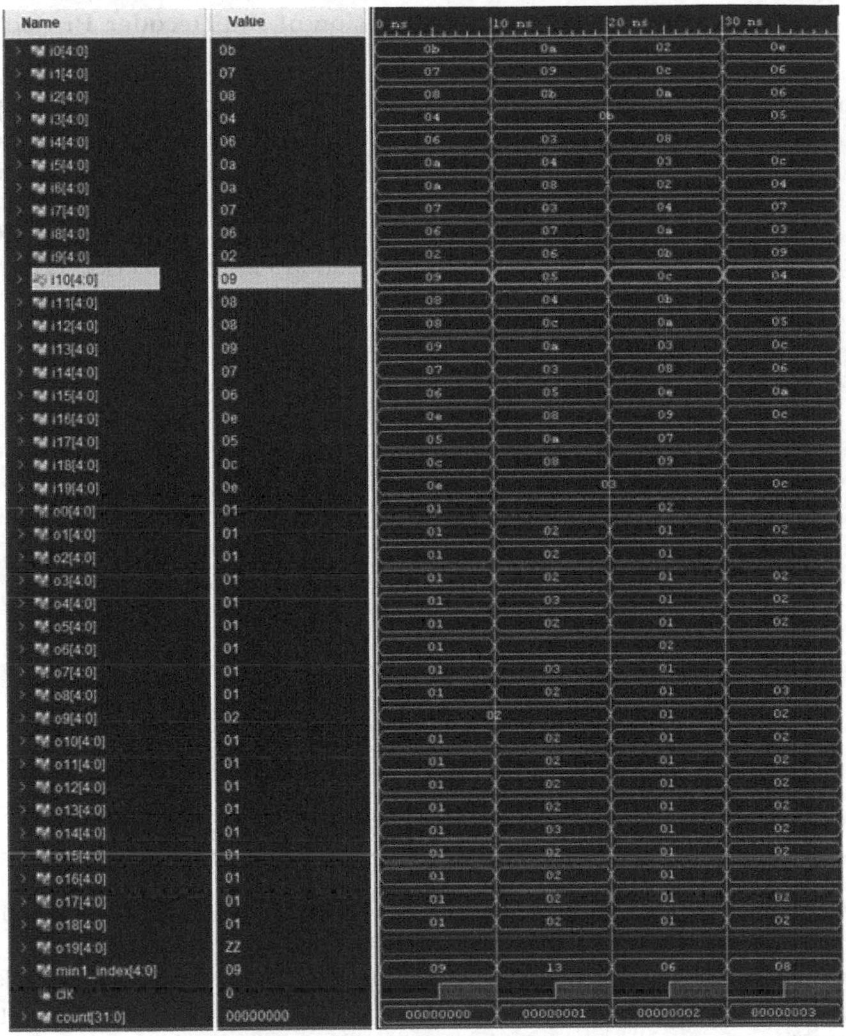

Name	Value	0 ns	10 ns	20 ns	30 ns
i0[4:0]	0b	0b	0a	02	0e
i1[4:0]	07	07	09	0c	06
i2[4:0]	08	08	0b	0a	06
i3[4:0]	04	04	0b		05
i4[4:0]	06	06	03	08	
i5[4:0]	0a	0a	04	03	0c
i6[4:0]	0a	0a	08	02	04
i7[4:0]	07	07	03	04	07
i8[4:0]	06	06	07	0a	03
i9[4:0]	02	02	06	0b	09
i10[4:0]	09	09	05	0c	04
i11[4:0]	08	08	04	0b	
i12[4:0]	08	08	0c	0a	05
i13[4:0]	09	09	0a	03	0c
i14[4:0]	07	07	03	08	06
i15[4:0]	06	06	05	0e	0a
i16[4:0]	0e	0e	08	09	0c
i17[4:0]	05	05	0a	07	
i18[4:0]	0c	0c	08	09	
i19[4:0]	0e	0e	03		0c
o0[4:0]	01	01		02	
o1[4:0]	01	01	02	01	02
o2[4:0]	01	01	02	01	
o3[4:0]	01	01	02	01	02
o4[4:0]	01	01	03	01	02
o5[4:0]	01	01	02	01	02
o6[4:0]	01	01		02	
o7[4:0]	01	01	03	01	
o8[4:0]	01	01	02	01	03
o9[4:0]	02	02		01	02
o10[4:0]	01	01	02	01	02
o11[4:0]	01	01	02	01	02
o12[4:0]	01	01	02	01	02
o13[4:0]	01	01	02	01	02
o14[4:0]	01	01	03	01	02
o15[4:0]	01	01	02	01	02
o16[4:0]	01	01	02	01	
o17[4:0]	01	01	02	01	02
o18[4:0]	01	01	02	01	02
o19[4:0]	ZZ				
min1_index[4:0]	09	09	13	06	08
clk	0				
count[31:0]	00000000	00000000	00000001	00000002	00000003

Fig. 10. Timing diagrams of LDPC decoder operation

In this test module, the input values i0,i1, ... i19 are generated using the built-in function $random. Since the $random function generates both positive and negative values, this function is preceded by another built-in function, $unsigned, which changes all values to unsigned. For better clarity of the algorithm, the value range was set from 2 to 15.

2.7   Results of Synthesis and Compilation of the Decoder Project

Based on the open source project [16], which is freely distributed, but has a number of significant limitations and disadvantages, an LDPC decoder project was created in the Xilinx Vivado FPGA development environment. Its design files and the digital circuit obtained as a result of synthesis are presented in Fig. 11.

The synthesized digital circuit was traced on an xczu3cg-sfvc784-2-i FPGA with 784 input-output elements, 70560 LUT, and 216 memory blocks. The tracing results are shown in Figs. 12, 13, and the involved resources of the xczu3cg-sfvc784-2-i FPGA are revealed in Table 3.

Table 3. FPGA xczu3cg-sfvc784-2-i resources used

Reference name	Involved	Functional category	Decoding
LUT5	198	LUT	lookup table
OBUF	100	IO	input/output interfaces
INBUF	100	I/O	input/output interfaces
IBUFCTRL	100	Other	-
LUT3	68	LUT	lookup table
LUT6	52	LUT	lookup table
LUT4	43	LUT	lookup table
OBUFT	5	IO	input/output interfaces
LUT2	1	LUT	lookup table

As can be seen from Table 3, the most required resources are related to LUT5 lookup tables. This is due to the fact that due to the specifics of the search for two minima, a large number of comparisons of elements are used. In second place are the input/output ports, which is caused by a fairly large number of input elements.

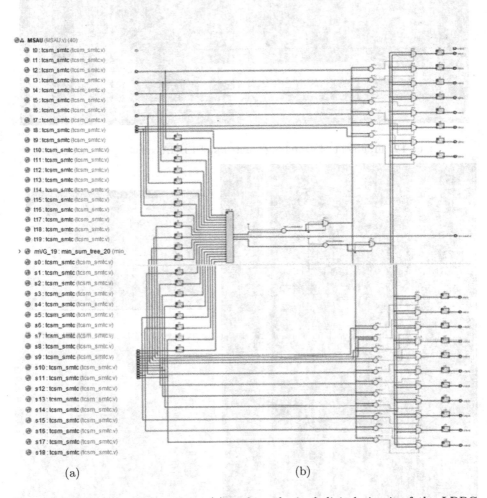

(a)   (b)

Fig. 11. Design files of test nodes (a) and synthesized digital circuit of the LDPC decoder (b)

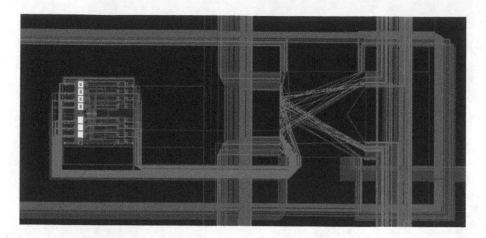

Fig. 12. Fragment of switching elements of FPGA xczu3cg-sfvc784-2-i

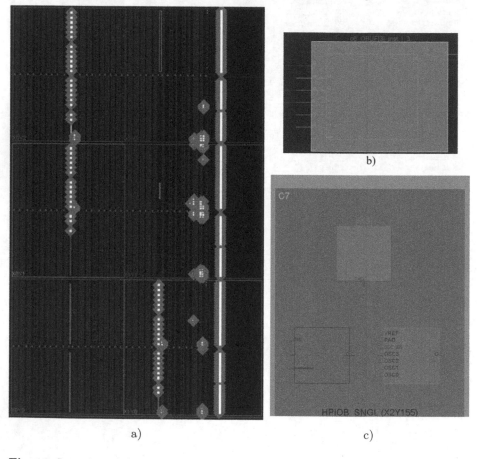

a)

b)

c)

Fig. 13. Location of the involved resources in the xczu3cg-sfvc784-2-i FPGA chip (a), functional cell (b), Input/Output block (c)

3 Conclusions

LDPC codes, used in the Wi-Fi, LTE and 5G mobile radio access network standards, are linear block codes that are determined by a sparse parity check matrix. The utilization of dedicated low-density parity check (LDPC) code for channel coding is an important technical advancement to ensure improved 5G NR performance compared to 4G and LTE. The minimum sum algorithm in the LDPC decoder is most suitable for implementation on field programmable logic devices [17].

Based on the specified detailed architecture of the LDPC decoder and analytical expressions describing the functioning of its blocks, the source code of the decoder project was created in the FPGA Verilog programming language. Test programs have been developed for the input and output TCSM/SMTC units, modules for finding the lesser of two and the upper level module of the LDPC decoder, which set a variety of test actions to check their functioning during simulation. The results of synthesis and tracing of the design in the selected FPGA confirm the possibility of its implementation in the selected chip or in a chip with comparable characteristics.

References

1. Detailed specifications of the terrestrial radio interfaces of International Mobile Telecommunications-2020 (IMT-2020), Rec. ITU-R M.2150-0, Geneva, Switzerland, February 2021
2. Aminev, D., Bogdanova, E., Kozyrev, D.: Analysis and formalization of requirements of URLLC, mMTC, eMBB scenarios for the physical and data link layers of a 5G mobile transport network. In: Vishnevskiy, V.M., et al. (Eds.) DCCN 2022, LNCS 13766, pp. 144–160. Springer, Cham (2022). https://doi.org/10.1007/978-3-031-23207-7_12
3. Technical Specification Group Radio Access Network; NR; Multiplexing and channel coding (Release 16), 3GPP TS 38.212 v16.4.0, December 2020
4. Li, Z., Chen, L., Zeng, L., Lin, S., Fong, W.H.: Efficient encoding of quasi-cyclic low-density parity-check codes. IEEE Trans. Commun. 54(1), 71–81 (2006)
5. Richardson, T.J., Urbanke, R.L.: Efficient encoding of low-density parity-check code. IEEE Trans. Inf. Theory 47(2), 638–656 (2001)
6. Jin, H., Khandekar, A., McEliece, R.J.: Irregular repeat accumulate codes. In: Proceedings of 2nd International Symposium Turbo Codes, Brest, France, pp. 1–8, September 2000
7. Ryan, W.E.: An introduction to LDPC codes Department of Electrical and Computer engineering, University of Arizona, August 2003
8. Petrovic, L., Markovic, M., El Mezeni, M., Saranovac, V., Radosevic, A.: Flexible high throughput QC-LDPC decoder with perfect pipeline conflicts resolution and efficient hardware utilization. Trans. Circ. Syst. I. 67(12), 5454–5467 (2020)
9. Gallager, R.: Low-density parity-check codes. IRE Trans. Inf. Theory. 8(1), 21–28 (1962)
10. Mackay, D.J.C., Neal, R.W.: Good codes based on very sparse matrices. In: Proceedings of 5th IMA Conference Cryptograph Coding, vol. 1025, pp. 100–111, December 1995

11. Verma, A., Shrestha, R.: A new VLSI architecture of next-generation QC-LDPC decoder for 5G new-radio wireless communication standard. In: International Symposium on Circuits and Systems, October 2020
12. Verma, A., Shrestha, R.: A new partially-parallel VLSI architecture of quasi-cyclic LDPC decoder for 5G new-radio. In: International Conference on VLSI Design, January 2020
13. Wey, C.-L., Shieh, M.-D., Lin, S.: Algorithms of finding the first two minimum values and their hardware implementation. IEEE Trans. Circuits Syst.–I Regular Papers. **55**(11), 3430–3437 (2008)
14. Yun, S., Kam, D., Choe, J., Kong, B.Y., Lee, Y.: Ultra-low-latency LDPC decoding architecture using reweighted offset min-sum algorithm. In: International Symposium on Circuits and Systems, October 2020
15. Kislitsyn, S.S.: On the selection of the k-th element of an ordered set by pairwise comparisons. Sibirskii Matematicheskii Zhurnal **5**(3), 557–564 (1964)
16. https://github.com/Abd1997-Dev/Design-of-Area-Efficient-Low-Latency-5G-Compliant-LDPC-Decoder-Architecture. Accessed 10 Feb 2024 — LDPC-Decoder project
17. Golovinov, E., Aminev, D., Tatunov, S., Polesskiy, S., Kozyrev, D.: Optimization of SPTA acquisition for a distributed communication network of weather stations. In: Vishnevskiy, V.M., Samouylov, K.E., Kozyrev, D.V. (Eds.) Distributed Computer and Communication Networks. DCCN 2020, LNCS, vol. 12563, pp. 666–679. Springer, Cham (2020). https://doi.org/10.1007/978-3-030-66471-8_51

Constructive Approach to Multi-position Passive Acoustic Localization in Information-Measurement Systems

Dmitry V. Churikov[1,2](✉) [ID]

[1] Scientific and Technological Centre of Unique Instrumentation of RAS,
Moscow, Russian Federation
churikov.d@gmail.com
[2] Kotelnikov Institute of Radioengineering and Electronics of RAS,
Moscow, Russian Federation

Abstract. The paper introduces a scheme for acoustic localization in the form of a one-dimensional "acoustic line", presenting the advantages and disadvantages of such a model. A constructive approach is proposed for multi-position passive acoustic location of sources in information-measuring systems. It is demonstrated that the application of R-functions and elements of contour analysis opens up additional opportunities for describing the consideration of parameters in the medium through which sound waves propagate. Experimental evidence shows that the use of planar and linear acoustic receiver systems enables the determination of the position of an impulse sound source with precision no worse than the measurement error. Special attention is devoted to advancing the method of passive acoustic localization by combining multiple approaches to enhance accuracy, reliability, computational speed, and efficiency.

Keywords: acoustic localization · digital signal processing · information measurement systems · multi-position methods · impulse signals · constructive approach · R-functions · contour analysis

1 Introduction

The research presented here builds upon previous work focusing on developing and implementing a method for multi-position passive acoustic localization of sound sources [1,2]. It has been demonstrated that in order for information systems to operate effectively, a multitude of parameters related to monitored processes must be taken into account. One crucial parameter is the timely detection not only of the occurrence of specific events, but also the determination of various characteristics such as the timing of events, their locations, conformity to predefined standards, frequency of occurrence, and more. Localization techniques can be categorized into passive (observational) and active (sensory) methods. Active methods offer advantages like the ability to detect objects that do

© The Author(s), under exclusive license to Springer Nature Switzerland AG 2024
V. M. Vishnevskiy et al. (Eds.): DCCN 2023, CCIS 2129, pp. 77–88, 2024.
https://doi.org/10.1007/978-3-031-61835-2_6

not emit sounds and the potential to enhance information gathering by adjusting the parameters of the probing signal. However, these systems come with the drawback of requiring energy expenditure during sensing, which can lead to unintended interference with other devices or components of the system.

On the other hand, passive methods, while being less sensitive, do not introduce additional impacts on the environment. By increasing the number of sensors and implementing new information processing approaches, we can enhance the accuracy and comprehensiveness of determining the parameters of observed objects. This is crucial for monitoring and controlling tasks in fields such as ecology, biomedicine, instrumentation, and mechanics. This work aims to improve the method of passive acoustic localization for model planar and linear systems. The advancement of this method includes investigating the characteristics of acoustic signals and evaluating the accuracy of sound source localization. The research also focuses on data processing processes using digital signal processing algorithms to enhance efficiency and computation speed. Combining different types of systems and approaches makes the method more versatile and effective for various application scenarios, including robotics control, event detection, as well as orientation and position control systems.

2   Problem Statement

Using the experimental setup [2], a study was conducted on the method of determining the position of a point sound impulse source based on signals from distributed acoustic sensors (receivers). Sound propagation in space is carried out by sound waves. Elastic waves are called mechanical disturbances (deformations) propagating in an elastic medium. The bodies that, acting on the environment, cause these disturbances are called wave sources. The propagation of elastic waves in the medium is not associated with the transfer of matter. In an unlimited medium, it consists in involving parts of the medium more and more distant from the source of the waves in forced oscillations. An elastic wave is longitudinal and is associated with volumetric deformation of an elastic medium, as a result of which it can propagate in any medium - solid, liquid and gaseous.

Table 1. Coordinates of the Receivers R_1, R_2, R_3 and source examples No. 1–4, absolute error is $\Delta = \pm 10$ mm.

Nodes	$\tilde{x}$	$\tilde{y}$
Receiver R_1	0	0
Receiver R_2	600	0
Receiver R_3	1200	0
Source position No. 1	300	250
Source position No. 2	750	250
Source position No. 3	600	500
Source position No. 4	1050	500

When an acoustic wave propagates in the air, its particles form an elastic wave and acquire an oscillatory motion, spreading in all directions if there are no obstacles in their path. The presence of obstacles in the way of propagation of sound waves affects both the propagating waves themselves and are sources of secondary waves. This is the cause of distortion, as well as the formation of an acoustic information leakage channel.

We make the following assumptions: the problem is considered on a plane, the sound wave is spherical, there are no reflections, the surrounding space is considered to be absolutely inelastic, the sensors and the sound source are considered to be point, and their characteristic size, i.e., $\Delta = \pm 10$ mm, is taken as the error in determining the coordinates.

In this paper, we will consider in more detail the case of placing receivers on the same line ("acoustic line"). This arrangement does not reduce the generality of the solution. In the second part of this work, this will reduce the task to one-dimensional and simplify the visualization of the proposed constructive approach for determining the position of the sound source. The experimental study was conducted for 4 positions of the sound source. The coordinates of acoustic sensors (receivers) and sound sources are presented in Table 1.

Figure 1 shows the geometry of the problem, where the signal from the source S with given coordinates propagates to three acoustic receivers R_1, R_3, and R_3 for example No. 4. The positioning accuracy is $\Delta = \pm 10$ mm.

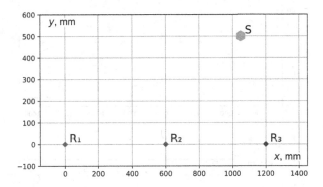

Fig. 1. The typical layout of the receivers R_1, R_2, and R_3 and the sound source S of the experimental installation for example No. 4 (Table 1).

3   Solution Method

As shown in [1,2], two linked sensors can form a stereo receiver. In this case, linkage is understood as the ability to receive time-synchronized signals from two sensors with negligible relative time delay. To determine the position of the sound source using multiple acoustic sensors, the method of two-dimensional acoustic location is used [1–7]. Figure 2 illustrates the path difference for the spherical sound wave from the source S to the receivers R_1, R_2, and R_3.

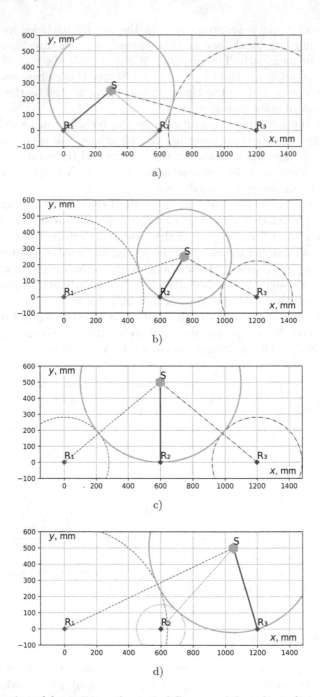

Fig. 2. Illustration of determining the path difference of the spherical sound wave from source S to receivers R_1, R_2, and R_3. Result for the source S positions No. 1 (a), No. 2 (b), No. 3 (c), and No. 4 (c) (Table 1).

This approach is based on measuring the time difference of arrival of sound signals to different sensors (acoustic path difference).

For the pair R_1 and R_2 let's determine the distance between the receivers (the length of section R_1R_2) and the time delay in registering the signal from the source to the receivers

$$\delta_{12} = t_1 - t_2. \tag{1}$$

Here t_1, t_2 are the time of signal registration by the receivers R_1 and R_2. Assuming the speed of sound in air V we can convert the time delay into distance

$$\Delta_{12} = V\delta_{12} = V(t_1 - t_2). \tag{2}$$

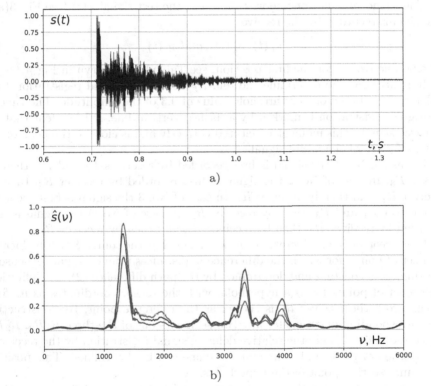

Fig. 3. The behavior of the test pulse signal $s(t)$ (a). Envelopes of signal spectra $\hat{s}(\nu)$ from receivers R_1 (red line), R_2 (green line), and R_3 (blue line) (b). (Color figure online)

4 Experiment

In the study, sound signals are captured using three acoustic sensors. The sampling frequency (44100 Hz) and quantization level (16 bits) were meticulously selected from the standard range.

These parameters are redundant for acoustic location purposes, but are necessary for a more detailed analysis of the approaches being developed. Upon reception, the captured signals undergo initial processing steps. This includes amplification and amplitude limitation to a level of 1. The settings for amplification and amplitude limitation play a crucial role in shaping the signals, thus impacting subsequent processing stages. While the specifics of these parameters necessitate a separate investigation beyond the scope of this work, for this experiment, the preliminary processing parameters are fine-tuned to ensure minimal distortion of the recorded signals during the rising interval. The Fig. 3 shows the behavior of the test pulse signal and envelopes of signal spectra from receivers R_1, R_2, and R_3.

The sound of a metronome was chosen as the test signal $s(t)$ (see Fig. 3(a)). The signals recorded by the receivers

$$s_1(t), s_2(t), \text{ and } s_3(t) \tag{3}$$

for example No. 1 (for the first position, see Table 1) are shown in Fig. 4.

It is necessary to determine the relative delay. The signal registration time is determined based on the threshold value of 0.3 of the amplitude. The time of the signal registration is marked by a dashed vertical line. It is evident that for all three signals, this moment is selected correctly and is close to the peak of the first half-wave of the useful signal.

In case of No. 1 the signal is first recorded by receivers R_1 and R_2, then by sensor R_3. In case of No. 2 the signal is first recorded by receiver R_3, then by receiver R_2, and then by receiver R_1. In case of No. 3 the signal is first recorded by receiver R_1 and R_3, then by receiver R_2. In case of No. 4 the signal is first recorded by receiver R_3, then by receiver R_2, and then by receiver R_1.

From geometric considerations, it is evident that the source S will be located on a set of points for which the difference in distances from each pair of sensors' coordinates is constant and determined by the path difference (2). By definition, such a set of points forms a hyperbola, with the sensor coordinates as its foci. In this work the receiver R_1 is chosen as the reference point. By constructing the corresponding diagrams for any two segments out of the three R_1R_2, R_2R_3, R_3R_1 and knowing only the relative delay in signal registration by the receivers, the position of the sound source on the plane can be determined. This position is the intersection point of the hyperbolas.

Results of locating the sound source position for two different positions (Table 1) are shown in Fig. 5. Enlarged fragments are presented in Fig. 5(b, d). Dashed lines represent numerical solutions, and a circle with a radius of $10\sqrt{2}$ mm illustrates the estimation area of calculation errors. The numerical results of determining the coordinates of the source for each of the variants are presented in Table 2. As seen from the obtained results, the error in determining the coordinates of the sources for the considered arrangements (Table 1) does not exceed the measurement error of the receiver coordinates. It should be noted that the results were obtained based on an approximate value of the speed of sound in air of $V = 345 \pm 5$ m/s. By reverse calculation, it is possible to experimentally determine the speed of sound based on the known source coordinates.

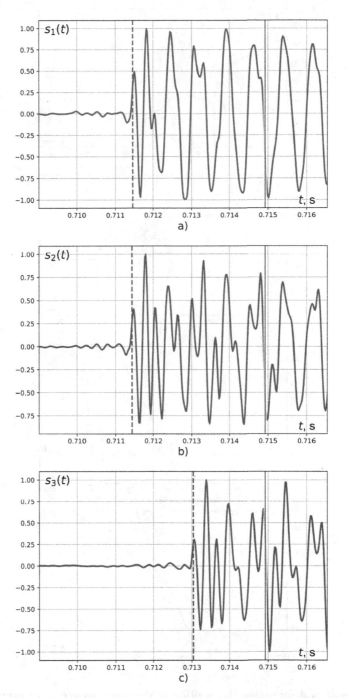

Fig. 4. The signals recorded by receivers R_1 (red line) (a), R_2 (green line) (b), and R_3 (blue line) (c) for the position of the source S No. 1 (Table 1). (Color figure online)

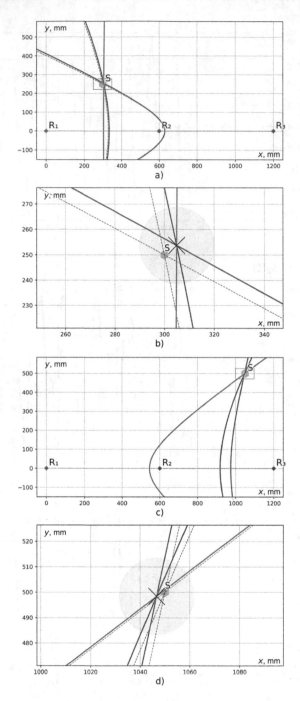

Fig. 5. Result of determining the position of the sound source S for examples No. 1 (coordinates is $(300; 250)$ mm) (a, b) and No. 4 (coordinates is $(1050; 500)$ mm) (c, d), $\Delta = \pm 10\sqrt{2}$ mm). General view (a, c), enlarged fragment (b, d). Red line for $R_1 R_2$, green line for $R_2 R_3$, blue line for $R_3 R_1$. Dashed lines are the numerical solutions. (Color figure online)

Table 2. Results of the experimental Source Coordinates determination.

Source example	x	$\Delta_x = |x - \tilde{x}|$	y	$\Delta_y = |y - \tilde{y}|$
No. 1	305	5	254	4
No. 2	752	2	254	4
No. 3	600	0	503	3
No. 4	1047	3	498	2

It can be seen from the results that the error in determining the position of the source only in one case exceeds the coordinate measurement error and does not exceed the estimate of the measurement error $\Delta_\Sigma = 10 \cdot \sqrt{2} \approx 14\,\text{mm}$.

5 Structural Approach for Multi-position Passive Acoustic Localization

Consider the case of a one-dimensional problem where 3 receivers and a source lie on the same straight line. In this case, the delay in recording the signal by each receiver will depend on only one x coordinate. Figure 6(a) shows an example for receivers R_1, R_2, and R_3 (Table 1) and the source S with the coordinates $(150; 0)$. The delay of signal registration in relation to the time of registration by receiver R_1 is postponed along the vertical axis. In the experiment, we assume that the position of the source S is unknown.

Considering that the sound wave from the source is spherical, we also assume that there are no obstacles in its propagation. Let's propose the following algorithm. On the plane $(x; \Delta t)$ we build lines connecting neighboring coordinates of points R_1R_2, R_2R_3 (see Fig. 6(b)):

$$k_{12} = \frac{\Delta t_2 - \Delta t_1}{x_2 - x_1}, \tag{4}$$

$$f_{12}(x) = k_{12}x + \frac{x_2\Delta t_1 - x_1\Delta t_2}{x_2 - x_1}, \tag{5}$$

$$k_{23} = \frac{\Delta t_3 - \Delta t_2}{x_3 - x_2}, \tag{6}$$

$$f_{23}(x) = k_{23}x + \frac{x_3\Delta t_2 - x_2\Delta t_3}{x_3 - x_2}. \tag{7}$$

Based on the assumptions made, it is obvious that the source coordinate will also lie on one or both of these lines (taking into account the error).

At the same time, if the lines, taking into account the error, coincide

$$k_{12} \approx k_{23} \text{ or } f_{12}(x) \approx f_{23}(x), \forall x, \tag{8}$$

Then the source coordinate will not lie inside the interval $x_S \notin (x_1, x_3)$. This case does not have a single solution.

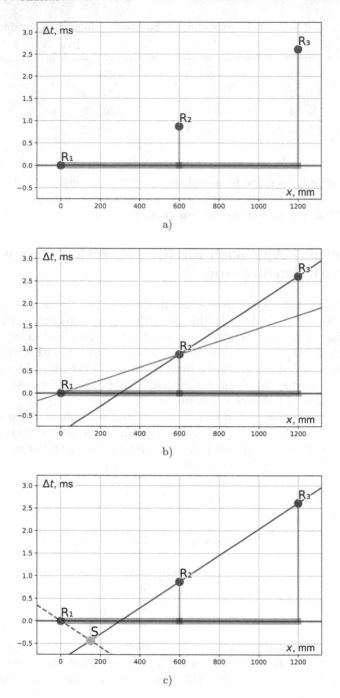

Fig. 6. Steps for determining the position of the sound source with the coordinate $(150; 0)$ mm: coordinates of receivers R_1, R_2, and R_3 on the plane $(x; \Delta t)$ (a), lines connecting neighboring coordinates of points $R_1 R_2$ (cyan line), $R_2 R_3$ (brown line) (b), the point of intersection of the lines defines the coordinate of the source (c). (Color figure online)

If they do not match $x_S \in (x_1, x_3)$. In this case, we determine the largest coefficient of the line

$$k = \begin{cases} -k_{12} \text{ for} & |k_{12}| > |k_{23}|; \\ -k_{23} & \text{otherwise.} \end{cases} \tag{9}$$

And plotting an additional line with a slope coefficient $-k$ passing through a point that does not lie on a straight line with a slope coefficient k (see Fig. 6(c)). Thus, the intersection of these two lines will determine the source coordinate $x_S \approx 149 \pm 2$ mm. And the slope coefficient will correspond to the inverse of the speed of sound in the medium $k \approx V^{-1}$.

If we apply a constructive approach [8], it is obvious that the position of the source S will be determined by the vertex of some inverted triangle. This triangle can be constructed as follows. Its base is strictly horizontal, the angles of inclination of the sides are equal to the angular coefficient k and on its sides, taking into account the error, all points lie all points R_1, R_2, R_3 and so on for more receivers in future work.

Thus, the solution of the problem is reduced to the construction of a triangle that meets the specified requirements.

Obviously, for the two-dimensional case, the solution also exists and can be found in the form of an inverted cone.

The constructive theory of R-functions [8] allows us to analytically define surfaces, including complex shapes. It is based on the use of Boolean operations to form a certain surface, which analytically defines the boundary of a complex object, the shape of the surface and its properties (boundary conditions).

This means that by registering signals with distributed receivers and knowing only the relative delays, it is possible to apply a constructive approach along with a computational approach. This together can improve the accuracy and speed of determining the coordinates of sound sources, including taking into account the anisotropy and non-stationarity of environmental parameters.

6   Conclusions

The primary objective of the study was to establish a functional experimental setup and verify fundamental relationships and development of a constructive approach for multi-position passive acoustic localization in information measurement systems. The research validates the capability of determining sound source coordinates utilizing receivers positioned arbitrarily with known coordinates at a precision level no less accurate than the coordinate measurement error.

These outcomes substantiate the viability of the proposed approach and the potential advancement of the methodology, particularly for exploring distant sources, discerning multiple concurrent signals, and incorporating environmental factors. The experimental findings exhibit consistency and reproducibility. Various series of measurements were executed during the preparation phase, aligning closely with the reference experiment outlined in the research.

Conducting the experiment in an acoustically untreated room yielded satisfactory results. Notably, signal shapes from different sensors demonstrate considerable synchronization within the delay determination zone. Distortions induced by environmental influences become apparent at a later stage.

This observation could prove valuable for investigating prolonged signals.

Future investigations will focus on exploring this correlation, optimizing parameters, evaluating assumptions' impact, examining conditions for implementing a streamlined solution framework, utilizing weighted processing methods and contour analysis techniques, and enhancing the efficiency of the devised approach further. Receiver placement can be facilitated through contour analysis and R-function [8] methodologies. This approach may enable the consideration of receiver and environmental characteristics, encompassing the medium's anisotropic and unsteady parameters.

The results were obtained using the equipment of the Centre for Collective Use of the Scientific and Technological Centre of Unique Instrumentation of the Russian Academy of Sciences (STC UI RAS) [456451, https://ckp.ntcup.ru/].

References

1. Churikov, D.V.: Method of multi-position passive acoustic location with distributed sensors. In: Proceedings of the XVI International Scientific and Technical Conference on Acousto-optical and Radar Methods of Measurements and Information Processing, Suzdal, Russia, 9–12 October 2023, pp. 72–75 (2023). https://doi.org/10.25210/armimp-2023-ABZBTY. EDN: ABZBTY, ISBN 978-5-6051133-3-1
2. Churikov, D.V.: Experimental study of the multi-position acoustic localization method for impulse sound source. In: Proceedings of the XXVI International Conference on Digital Signal Processing and Its Applications (DSPA), Moscow, Russia, 27–29 March 2024, pp. 1–4 (2024). https://doi.org/10.1109/DSPA60853.2024.10510027
3. Rafaely, B.: Fundamentals of Spherical Array Processing, 2nd edn. Springer, Cham (2019). https://doi.org/10.1007/978-3-319-99561-8, ISBN 978-3-319-99560-1
4. Stolbov, M.B.: Application of microphone arrays for remote speech information gathering. Sci. Tech. Bull. Inf. Technol. Mech. Opt. 15(4), 661–675 (2015)
5. Hudson, R., Yao, K., Chen, J.: Acoustic source localization and beamforming: theory and practice. EURASIP J. Adv. Sig. Process. 2003 (2003). https://doi.org/10.1155/S1110865703212038
6. Benesty, J., Huang, G., Chen, J., Pan, N.: Microphone Arrays. Springer, Cham (2024). https://doi.org/10.1007/978-3-03136974-2
7. Ajdler, T., Kozintsev, I., Lienhart, R., Vetterli, M.: Acoustic source localization in distributed sensor networks. In: Proceedings of the 38th Asilomar Conference on Signals, Systems and Computers. IEEE (2004). https://doi.org/10.1109/acssc.2004.1399368
8. Kravchenko, V.F., Churikov, D.V.: Digital signal processing by atomic functions and wavelets. J. Commun. Technol. Electron. 68(suppl. 1), S1–S110 (2023). https://doi.org/10.1134/S1064226923130016, ISSN 1064-2269

Application of Queueing Theory to Investigation of HaProxy Load Balancer Performance Characteristics

Aleksandr Sokolov[✉][iD]

V.A. Trapeznikov Institute of Control Sciences Russian Academy of Sciences,
Profsoyuznaya Street 65, 117997 Moscow, Russia
aleksandr.sokolov@phystech.edu

Abstract. This paper explores the application of queuing theory to assess the performance characteristics of the HaProxy load balancer, including response time, service waiting time, and the average number of busy servers. To approximate the load balancing process, we used $MMAP/PH/M/N$ queuing analytical model type with a Marked Markovian input flow, the service time has PH - distribution, the system has a buffer of M capacity, the number of serving devices equals N. We use discrete event simulation (Monte Carlo method) to estimate the performance characteristics to simplify the process and save time instead of using the complex and time-consuming analytical calculations. In the paper, a numerical experiment was carried out to compare the performance values obtained through simulation with the metrics obtained from the test server. The effectiveness of the application of models was demonstrated.

Keywords: queuing theory · load balancing · priority queue

1 Introduction

Load balancing is a part of any extensive web application. The essence of the load balancer is to distribute the network load on the system between several servers. At the start of the web application, it may occupy few resources on a single virtual or physical machine. The server's characteristics could be improved, such as the number of CPU cores, RAM, and disk space, to scale the application. Later on, when scaling the system, vertical scaling leads to higher costs than the resulting gain in system performance. Due to the increasing load on the application, horizontal scaling with load balancing is used for greater fault tolerance. A balancer in the system evenly distributes the network load between servers. This paper explores the feasibility of applying the queuing theory to derive performance estimates for a system with load balancer that handles prioritized traffic.

The objectives of balancing are:

The research was funded by the Russian Science Foundation, project no. 22-49-02023.

- reduction of system response time is the time from sending a request by a client to receiving a response;
- increasing system fault tolerance, i.e. if at least one server in the system fails, the system will still continue to operate;
- efficient resource allocation, where the system's servers receive approximately the same load.

There are a large number of different load balancing technologies and algorithms. The article [3] provides a comprehensive overview of balancing algorithms. The article [9] investigates the effectiveness of various balancing algorithms when using HaProxy as a load balancer. The authors of the article [14] investigate the effectiveness of using HaProxy load balancer against Dos attacks. The experiments were conducted on a variety of OS, including Linux and Windows. A dynamic balancing algorithm is proposed in [6]. Propose an algorithm to estimate server utilization using CPU utilization and communication link bandwidth between the balancer and the server. The algorithm's efficiency has been demonstrated, surpassing that of the classical Weighted Round-robin algorithm. The paper [7,8,11] investigates the efficiency of various balancing algorithms, including Round-robin, Least Connections, and others.

The performance characteristics of various balancing technologies were thoroughly investigated. The paper [8] analyzes and contrasts the performance of the HaProxy and Nginx load balancers, specifically evaluating their throughput, response time, and other characteristics. This study examines the effectiveness of utilizing a load balancing system in comparison to a system without load balancing. In the paper [4] thoroughly investigated the efficiency of load balancing in SDN networks.

Queueing theory is also applied to study the efficiency of load balancers and realize algorithms for dynamic load balancing among servers [10,13,15]. An exemplary balancing algorithm is presented in [13]. In order to operate the algorithm, it is necessary to set the average service intensity μ_i for each server, the request arrival intensity λ, and the maximum queue size of requests on each server. Based on these parameters, a formula is proposed for calculating the weights for each server when applying Weighted Round-robin balancing. An analytical model for the study of load balancing is given in the paper [15]. The paper proposes the implementation of an infinite buffer in the load balancer. After the load balancer, the request is queued for service to a free server. The server is denoted by $M/M/1/K$ in Kendall's terms, the input is Poisson flow, the service time is exponentially distributed, and the maximum number of requests in the queue is K. The paper presents a variety of balancing schemes and their corresponding Markov chain transitions. It provides formulas for analyzing the performance characteristics of the system.

This paper explores the feasibility of applying the $MMAP/PH/M/N$ model [5,12] to derive performance estimates for a load-balanced system that handles prioritized traffic. The service time is modeled by a PH - distribution, the system has a buffer capacity of M, the number of servers in the system is N, and the system includes K priority classes. In this paper, we conducted a numerical experiment to demonstrate how the model can be used to assess the performance of the balancer and the cluster. Estimates obtained with the help of queueing theory lie within the acceptable error. This method can be used in designing web applications.

The article is divided into three sections. The first section discusses different types of load balancing and briefly introduces the popular HaProxy technology. The second section provides an analytical model of queueing theory that can be used to estimate the performance characteristics of the HaProxy load balancer. Finally, the third section presents the results of a numerical experiment that compares the results obtained from the HaProxy test server and simulation modeling.

2 Description of Load Balancing Mechanisms. HaProxy Load Balancer

This section summarizes the types of load balancing and provides a brief description of HaProxy technology. The most commonly used load balancers in practice are L4 and L7.

L4 balancing (Fig. 1) is the simplest method for balancing network load in a system. This balancer works at the transport layer of the OSI model protocol stack. In this case, packet transmission occurs as follows. The client connects to the load balancer, which then connects to one of the servers and transfers data from the client to the server. After completing the balancer-server TCP session, the balancer responds to the client-balancer TCP session. Proxying at this layer is easy and fast. This balancing method offers multiple features. It can perform health checks, hide the internal network used for the balancer-server connection, queue connections to prevent server overload, and limit connection speeds. This type of balancing is suitable for web applications working with databases such as PostgreSQL, Redis, etc.

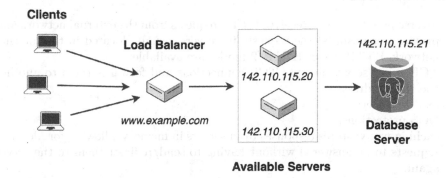

Fig. 1. L4 load balancing scheme.

In L7 (Fig. 2), balancing occurs at OSI layer 7, the application layer. With this method, the balancer can make routing decisions based on HTTP request data, such as source IP addresses and HTTP metadata: headers, cookies, URLs, etc.

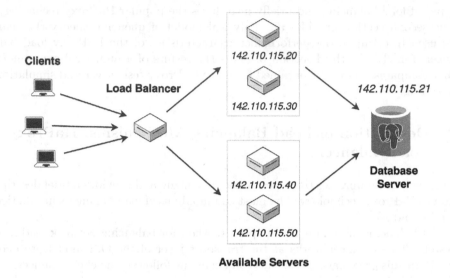

Fig. 2. L7 load balancing scheme.

2.1 HaProxy Load Balancer

One of the popular technologies used for load balancing is HaProxy, which is widely utilized by major web services and platforms such as GitHub, Stack Overflow, Reddit, AWS, and others. HaProxy technology can perform many functions, including:

- reverse proxy server: it receives HTTP requests from the external network and proxies (retransmits) the requests to servers, usually located in the internal network. HTTP/1.x or HTTP/2 modes are available;
- TCP proxy server: accepts TCP connections and forwards them to another specified address;
- load balancer;
- DOS protections;
- caching proxy: stores responses from servers in memory, allowing for identical requests to be answered without having to send/redirect them to the server again.

HaProxy uses a non-blocking I/O architecture with a multi-threaded scheduler. Its architecture is designed to move data as quickly as possible with the fewest number of operations. HaProxy optimizes CPU cache handling by running a single connection on a single core, eliminating the need for switching. During TCP data proxying, 15% of the time is spent on HaProxy operations, while 85% of the time is spent on operating system operations, such as opening/closing the connection and forwarding data.

Only a configuration file is required to use HaProxy. After launch, HaProxy balancer performs the following functions: accepting incoming connections, periodic server health checks, and communication with other HaProxy instances.

2.2   Queues and Priorities in HaProxy

To prevent server crashes caused by memory overflow, HaProxy limits active connections. To limit the number of connections, you can make use of the *max-conn* directive. Depending on the server's resources and the difficulty of the tasks being performed, it's important to select an appropriate value for the maxconn parameter to prevent the server from overloading and failing. Additionally, the configuration allows you to specify the number of requests that can be queued in case of busy or failed servers. This parameter can be set using the *timeout queue* directive, where the time is specified in seconds. When a client's request is in the queue for the specified time, and all servers are busy or fail, the client will receive a response with the code 503 (Service Unavailable).

```
1  # Configuration of frontend server
2  frontend myfrontend
3      bind :80
4      maxconn 1000
5      default_backend webservers
6
7  # Configuration of backend services
8  backend webservers
9      # Round-robin-balancing
10     balance roundrobin
11     acl is_priority path_beg /priority/
12     http-request set-priority-class int(1) if is_priority
13     http-request set-priority-class int(2) if !is_priority
14     timeout queue 30s
15     server s1 192.168.110.10:80 maxconn 100
16     server s2 192.168.110.11:80 maxconn 30
17     server s3 192.168.110.12:80 maxconn 50
```

Listing 1.1. Configuring Priority Access in HaProxy

HaProxy 1.9 has introduced a new feature for traffic prioritization. To prioritize traffic, you need to define certain rules in the configuration file, known as ACLs (Access Control Lists, https://www.haproxy.com/documentation/haproxy-configuration-tutorials/core-concepts/acls/). ACLs help you set rules for incoming requests, such as determining the servers where traffic can be accessed or forwarded, and determining the client priority, among others.

The example in listing 1.1 shows that the rule *is_priority* is assigned a value of *true* if the path starts with */priority/*, and *false* otherwise. After that, the traffic is assigned a corresponding priority. It's worth noting that priority requests are queued ahead of non-priority requests, and priority clients are served first.

3   A Prioritized Balancing Model. Description of the Analytical Model $MMAP/PH/M/N$

The HaProxy balancing system can be represented schematically as follows. The frontend manages a queue of connections, prioritizing them based on creation time or set priorities. The maximum number of active connections allowed can be defined for the frontend in the configuration file using the *maxconn* directive. Similarly, the *maxconn* directive can also be used to specify the number of simultaneous connections allowed for each backend server. Besides, the system allows configuring different timeouts such as the maximum time required to establish a connection between the client and frontend, the maximum server response time, the maximum time required to queue a request, and other time settings.

The diagram in Fig. 3 shows how load balancing works in HaProxy technology. When an HTTP request arrives, it is stored in a prioritized queue for a

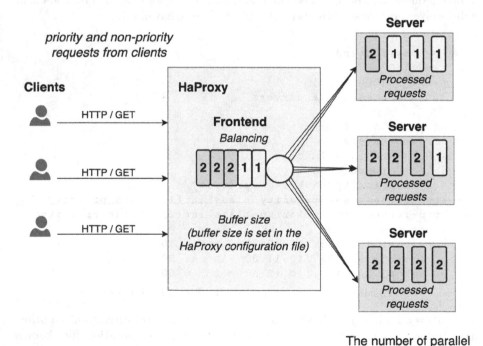

Fig. 3. HaProxy balancing scheme.

designated time, as specified in the configuration file. If the request is not forwarded to the server within the specified time, the connection to the client is dropped. In addition, each server has a defined number of simultaneous connections.

3.1 $MMAP/PH/M/N$ Queuing Model

The system shown in Fig. 4 is a queuing system with priority service. It receives customers in the form of a Marked Markovian Arrival Process (MMAP) [1,2] and has K-priority customers. A buffer of limited size M is present in the system, where priority customers are given preference over the lower priority customers. The system has N servers, and each priority customer is serviced according to a service time distribution PH_i, where $1 \leq i \leq K$.

In the paper [12], a two-priority system ($K = 2$) is studied, and stationary probabilities are obtained. The paper provides formulas for finding the performance characteristics of the system. However, it is shown that the computation time of the stationary states grows exponentially with the system parameters, making it computationally expensive. Therefore, in order to solve the general problem where K is arbitrary, the discrete event modeling method is applied.

From the description, we can see that an analogy can be drawn between the HaProxy balancing system model and the $MMAP/PH/M/N$ model. Client requests or connections to the frontend server are analogous to customers in the priority system, the queue of requests (connections) in the frontend server is analogous to the queue of customers in the priority system, and the total number of possible connections on all servers is analogous to servers of queuing system.

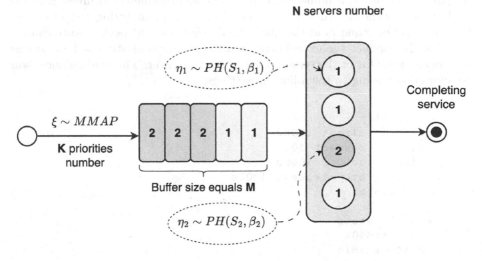

Fig. 4. Queuing system $MMAP/PH/M/N$.

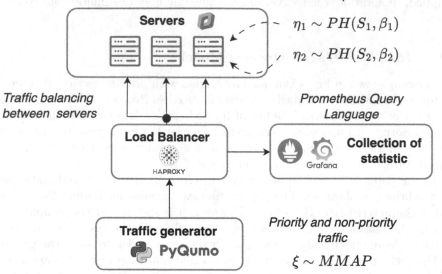

Fig. 5. An design og experiment is being developed to study the performance characteristics of the HaProxy balancer.

4   Numerical Results

As part of this work, a numerical experiment was conducted to investigate the performance of the HaProxy system. The focus was on analyzing response time, waiting time for requests in the queue, and other relevant performance characteristics. To conduct the experiment, a small cluster consisting of four servers was assembled. One of the servers used as load balancer. The load balancer was configured according to the following settings:

```
1    defaults
2        mode http
3        timeout client 1000s
4        timeout connect 10s
5        timeout server 1000s
6        timeout http-request 1000s
7        log global
8
9    frontend stats
10       bind *:8404
11       stats enable
12       stats uri /
13       stats refresh 10s
14
```

```
15    frontend prometheus
16       bind *:8405
17       mode http
18       http-request use-service prometheus-exporter if { path
         /metrics }
19       no log
20
21    frontend myfrontend
22       bind :80
23       maxconn 1000
24       default_backend webservers
25
26    backend webservers
27       acl is_checkout path_beg /priority
28       http-request set-priority-class int(1) if is_checkout
29       http-request set-priority-class int(2) if !is_checkout
30       server s1 web1:5010 maxconn 30
31       server s2 web2:5010 maxconn 30
32       server s3 web3:5010 maxconn 30
```

Listing 1.2. HaProxy configuration

To minimize network latency, virtual machines were rented from the same cloud for the servers and the balancer, as shown in Fig. 5. For the experiment, four virtual machines were rented, each with an Intel Ice Lake platform and an Intel Xeon Gold 6338 processor with a clock frequency of 2.00 GHz, as well as 2 GB of RAM. Each virtual machine runs an http server that accepts GET requests for both priority and non-priority clients. The request execution time for both types of users was assumed to be exponentially distributed, and the average request processing time for priority and non-priority users was set. To simulate long request processing on the server, the function $sleep(n)$ was executed, where n is the number of seconds for which the request execution stopped, distributed according to some specified distribution. To generate client traffic, a Python script was used, which was written using the asyncio, requests_async libraries to organize asynchronous requests, and the pyqumo library (https://github. com/ipu69/pyqumo) to generate time intervals between consecutive requests. HaProxy collects Prometheus metrics out of the box. The Grafana service was used to aggregate and display statistics. Virtual machines for the servers and for the balancer were rented in the same cloud to minimize network latency (Fig. 5).

The experiment used the configuration listed in Listing 1.2. Priority and non-priority requests were distributed exponentially, with an average processing time of three seconds for both types of requests ($\mu_{pr} = \mu_{npr} = \mu$). Each server could handle a maximum of 30 queries at a time. The input was a stream called $MMAP$, with equal arrival intensity for both priority and non-priority requests ($\lambda_{pr} = \lambda_{npr} = \lambda$). The experiment aimed to determine how different performance characteristics (such as system response time for priority and non-priority requests, queue time, and number of active connections) were affected by the system utilization factor. The results were compared with those obtained through simulation modeling.

The results of the experiment are presented in Fig. 6. To obtain the performance characteristics for each load factor, we generated 100,000 queries and ran each test eight times. The chart also displays the measurement errors. The plots indicate that the simulation results align with the metrics obtained from the HaProxy system within the error bounds. The measurement error increases with an increase in the loading factor. For instance, at $\rho \geq 0.9$, the error in response time is around 15%, while for waiting time for non-priority request packets, it is about 20%. The error in other characteristics was not more than 10%. The comparison of the number of active connections of the balancer to the servers is shown in Fig. 7. The simulation results align with the system metrics within the margin of error.

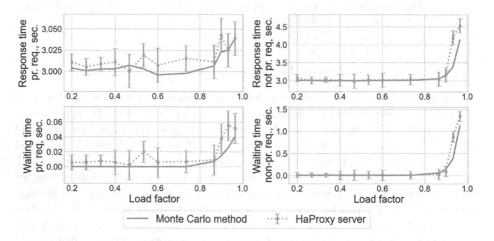

Fig. 6. Comparison of HaProxy balancer performance characteristics with simulation (Monte Carlo) results for priority and non-priority requests.

Getting metrics from the test server is a time-consuming process. As shown in the Table 1, it can take up to several hours to obtain stable characteristics, depending on the system load factor. However, it takes no more than a minute to obtain stable estimates using the Monte Carlo method with 1,000,000 simulated customers. You can find all the results of the study in the Github repository at the following URL: https://github.com/timac11/thesis-queuing-models-examples.

Table 1. Average time to get performance characteristics depending on load factor.

Method	$\rho = 0.2$	$\rho = 0.4$	$\rho = 0.6$	$\rho = 0.8$	$\rho = 0.9$
HaProxy Server	35 h	16 h	12 h	8 h	6 h
Monte Carlo method	10 s	15 s	24 s	32 s	38 s

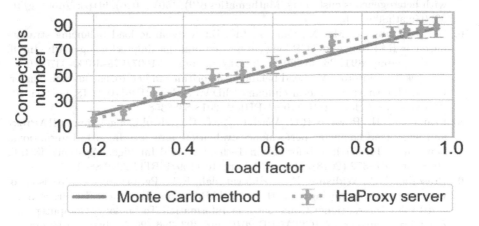

Fig. 7. Comparison of HaProxy balancer busy active connections number with simulation (Monte Carlo) results for priority and non-priority requests.

5 Conclusion

The purpose of our study was to explore the feasibility of applying queueing theory to analyze the performance of a load balancer. We've discovered that a model with a correlated input $MMAP$ flow of the form $MMAP/PH/M/N$ can be used to study traffic, including prioritized traffic. In this model, the service time follows a PH distribution, the buffer size is M, and there are N servers. This method of evaluating performance characteristics can be incredibly useful in the design and load estimation of large web applications. It has advantages over traditional testing methods, such as saving time and reducing costs associated with renting and configuring servers.

References

1. Bocharov, P.P., D'Apice, C., Pechinkin, A.V.: Queueing Theory. De Gruyter, Boston (2003). https://doi.org/10.1515/9783110936025
2. Dudin, A.N., Klimenok, V.I., Vishnevsky, V.M.: The Theory of Queuing Systems with Correlated Flows, 1st edn. Springer, Cham (2019). https://doi.org/10.1007/978-3-030-32072-0
3. Jader, O.H., Zeebaree, S.R., Zebari, R.R.: A state of art survey for web server performance measurement and load balancing mechanisms. Int. J. Sci. Technol. Res. 8(12), 535–543 (2019)

4. Khaliq, A., Tahir, M.A., Nadeem, G., Adil, S.H., Jamshid, J., Memon, J.A.: Performance comparison of webservers load balancing using HAProxy in SDN. In: 2023 4th International Conference on Computing, Mathematics and Engineering Technologies: Sustainable Technologies for Socio-economic Development, iCoMET 2023, pp. 1–5 (2023). https://doi.org/10.1109/iCoMET57998.2023.10099326
5. Klimenok, V., Dudin, A., Vishnevsky, V.: Priority multi-server queueing system with heterogeneous customers. Mathematics 8(9), 1501 (2020). https://doi.org/10.3390/math8091501
6. Li, W., Liang, J., Ma, X., Qin, B., Liu, B.: A dynamic load balancing strategy based on HAProxy and TCP long connection multiplexing technology. Adv. Intell. Syst. Comput. 891, 36–43 (2019). https://doi.org/10.1007/978-3-030-03766-6_5
7. Mbarek, F., Mosorov, V.: Load balancing algorithms in heterogeneous web cluster. In: 2018 International Interdisciplinary PhD Workshop, IIPhDW 2018, pp. 205–208 (2018). https://doi.org/10.1109/IIPHDW.2018.8388358
8. Pramono, L.H., Buwono, R.C., Waskito, Y.G.: Round-robin algorithm in HAProxy and nginx load balancing performance evaluation: a review. In: 2018 International Seminar on Research of Information Technology and Intelligent Systems, ISRITI 2018, pp. 367–372 (2018). https://doi.org/10.1109/ISRITI.2018.8864455
9. Prasetijo, A.B., Widianto, E.D., Hidayatullah, E.T.: Performance comparisons of web server load balancing algorithms on HAProxy and Heartbeat. In: Proceedings - 2016 3rd International Conference on Information Technology, Computer, and Electrical Engineering, ICITACEE 2016, pp. 393–396 (2017). https://doi.org/10.1109/ICITACEE.2016.7892478
10. Proskochylo, A., Zriakhov, M., Akulynichev, A.: The effects of queueing algorithms on QoS for real-time traffic in process of load balancing. In: 2018 International Scientific-Practical Conference on Problems of Infocommunications Science and Technology, PIC S and T 2018 - Proceedings, pp. 575–580 (2018). https://doi.org/10.1109/INFOCOMMST.2018.8632161
11. Saxena, P.K.: Mohit: a round-robin based load balancing approach for scalable demands and maximized resource availability. Int. J. Eng. Comput. Sci. 5(8), 17375–17380 (2016)
12. Vishnevsky, V., Klimenok, V., Sokolov, A., Larionov, A.: Performance evaluation of the priority multi-server system MMAP/PH/M/N using machine learning methods. Mathematics 9(24), 3236 (2021). https://doi.org/10.3390/math9243236
13. Wen, Z., Li, G., Yang, G.: Research and realization of nginx-based dynamic feedback load balancing algorithm. In: Proceedings of 2018 IEEE 3rd Advanced Information Technology, Electronic and Automation Control Conference, IAEAC 2018, pp. 2541–2546 (2018). https://doi.org/10.1109/IAEAC.2018.8577911
14. Zeebaree, S.R., Jacksi, K., Zebari, R.R.: Impact analysis of SYN flood DDoS attack on HAProxy and NLB cluster-based web servers. Indones. J. Electr. Eng. Comput. Sci. 19(1), 505–512 (2020). https://doi.org/10.11591/ijeecs.v19.i1.pp505-512
15. Zhang, Z., Fan, W.: Web server load balancing: a queueing analysis. Eur. J. Oper. Res. 186(2), 681–693 (2008). https://doi.org/10.1016/j.ejor.2007.02.011

Broadband Wireless Networks Based on Tethered High-Altitude Unmanned Platforms

Vladimir Vishnevsky[1]($\boxtimes$) , Yuriy Avramenko[1] , Van Hieu Nguyen[2] ,
and Nikita Kalmykov[1]

[1] Institute of Control Sciences of Russian Academy of Sciences, Profsoyuznaya Street
65, Moscow 117997, Russia
vishn@inbox.ru, avramenko@ipu.ru
[2] Moscow Institute of Physics and Technology, Institutsky lane 9, Dolgoprudny,
Moscow Region 141700, Russia
hieu.nguyen@phystech.edu

Abstract. This paper describes the advantages of implementing a broadband wireless network based on a tethered drone and assessing its performance characteristics. The paper presents the calculation results for the increase in the telecommunications coverage area (line-of-sight zone) and the parameters of the communication channel between the base station (BS) located on the drone and the ground station (GS) within line-of-sight. A stochastic polling model with batch packet servicing is proposed to evaluate the network performance. A description is given of the interaction protocol between the BS and the GS for obtaining initial data when carrying out numerical calculations using a simulation model that adequately describes the functioning of a broadband wireless network with a centralized control policy.

Keywords: tethered drone · wireless network · line of sight · stochastic polling

1 Introduction

Currently, autonomous unmanned aerial vehicles (UAVs) have found wide application in both civil and defense industries. They are used especially effectively to create modern broadband wireless networks and various communication systems [2,15–18]. The main disadvantage of autonomous UAVs and, accordingly, telecommunication networks based on them is the limited operating time (within 1 h), which is associated with the short life of the battery installed onboard the UAV. In this regard, such UAVs cannot be effectively used in systems that require a long operating time, for example, in security management systems and

This study was carried out with financial support from the Russian Science Foundation, grant No. 22-49-02023.

protection of critical facilities (nuclear power plants, airfields, long bridges and state border sections).

Long-term operation can be ensured by tethered UAVs (tethered high-altitude unmanned platforms). The power supply of their propulsion systems and payload equipment is carried out from ground-based energy sources via a tether cable. Tethered high-altitude unmanned platforms occupy an intermediate position between satellite systems and ground-based systems, the equipment of which (cellular base stations, radio relay and radar equipment, etc.) is located on high-rise structures. Compared to expensive satellite systems, tethered UAVs are highly cost-effective, and surpass the ground-based telecommunications systems in terms of telecommunications and video coverage. The latest reviews [3,10] on the topic of tethered high-altitude platforms provide links to numerous articles on design methods and architecture of such platforms based on high-altitude balloons, airships and aircrafts. However, the topic of tethered high-altitude unmanned platforms, which use multi-rotor unmanned aerial vehicles (tethered drones) as a high-altitude module, is poorly covered. Tethered drones (tethered UAV or tUAV), the intensive development of which began in the last decade, are capable of solving many new problems in both the civil and defense industries [5,6,9,11,12]. The advantage of such networks is fast and flexible deployment, increased reliability of wireless communications, controlled mobility, reduced operating costs, etc. The design and implementation of broadband wireless networks based on tethered drones is a promising direction in the creation of next generation 5G/6G networks.

This paper evaluates the performance of a broadband wireless network based on a tethered drone using an original stochastic polling model. In contrast to well-known works [4,13,14], we consider a polling model with batch servicing of queues of packets from wireless network subscribers. The calculation of the maximum radio coverage, required power, and antenna gain factors for receiving and transmitting wireless network devices based on a tethered drone is also provided.

2   Calculation of the Telecommunications Coverage Area and Characteristics of the Commu-Nication Channel of a Wireless Network Based on a Tethered High-Altitude Unmanned Platform

Tethered drones of the new generation provide the ability to lift a telecommunications payload of significant weight (up to 10–15 kg) and long-term operation at altitudes up to 100–150 m. The following can be used as a payload: communications equipment in the decimeter range of radio waves for organizing communications with ground or air users at a distance of 100 km or more; a cellular base station for the rapid creation of a modern telecommunications infrastructure over a vast territory or network equipment operating under the control of IEEE 802.11 protocols in point-to-multipoint mode in the centimeter wave range. Effective operation of telecommunications equipment in the specified frequency ranges is ensured by increasing the line-of-sight zone between the antennas of receiving and transmitting equipment and reducing the level of interference.

The maximum communication range, depending on the lifting height of the tethered drone, practically coincides with the distance to the visible horizon (Fig. 1).

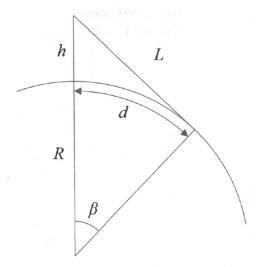

Fig. 1. Line of sight area

Here h is the height of the BS placement, L is the maximum communication range, d is the length of the arc along the Earth's surface between the projection of the copter onto the surface and the line limiting the line of sight zone.

The indicated distance is determined from the following simple geometric relationships:

$$d = R \ \arctan\left(\sqrt{\left(1+\frac{h}{R}\right)^2 - 1}\right) \ ; \ L = \sqrt{h^2 + 2Rh}, \tag{1}$$

Assuming that the shape of the earth is an ideal sphere of radius $R \approx 6.37 \times 10^6$ m, we obtain the following data on the size of the line of sight zone (Table 1, Fig. 2).

Thus, the utilization of high-altitude unmanned tethered platforms, with a lifting height of 100–150 m, allows us to obtain a line of sight zone of 35–43 km. This calculation was made for the case when the receiving equipment antennas are at ground level.

For the effective functioning of a communication channel within the line-of-sight zone calculated above (telecommunications coverage area), it is necessary to estimate the amount of transmitter power, the amount of losses in the communication line, and the gains of the receiving and transmitting antennas. The

Table 1. Line of sight size

h, m	l, m	d, m
1	3569	3569
10	11287	11287
50	25239	25239
100	35693	35693
150	43715	43715

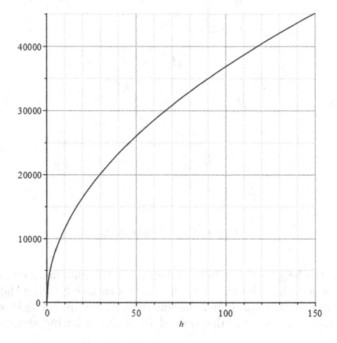

Fig. 2. Dependence of the line of sight zone on the lifting height of the tethered drone

power received at the receiver input can be determined using the Friis formula [7]. A simplified version of this formula is as follows:

$$P_r = P_t G_t G_r \left(\frac{\lambda}{4\pi L} \right)^2, \qquad (2)$$

where P_t is the transmitter power, G_t and G_r are the gains of the transmitting and receiving antennas, λ is the wavelength and L is the distance between the receiving and transmitting antennas. In formula (2), antenna gains are dimensionless quantities, and the units of wavelength (λ) and distance (L) must be the same. The main reason for "losses" is the reduction in signal power due to the uniform propagation of radio waves. Signal attenuation is proportional to the square of the distance. By taking logarithm of (2), the Friis formula can be transformed [8] to the following form:

$$P_r = P_t + G_t + G_r - \Pi, \tag{3}$$

where P_r and P_t are expressed in decibels relative to milliwatts, and the gains of the transmitting and receiving antennas G_t and G_r are expressed in decibels relative to the isotropic emitter. Parameter Π (4) determines the attenuation in the wireless line:

$$\Pi = 20 \times \log_{10}(4\pi\frac{L}{\lambda}), \tag{4}$$

where λ is the wavelength and L — the distance between the receiving and transmitting antennas, expressed in the same units.

Table 2 shows the attenuation values Π depending on the distance between the receiver and transmitter for various frequency ranges.

Table 2. Signal attenuation

Frequency, MHz	Attenuation Π, dB		
	25 km	35 km	43 km
100	100.41	103.33	105.12
200	106.43	109.35	111.14
400	112.45	115.37	117.16
800	118.47	121.39	123.18
1000	120.41	123.33	125.12
2000	126.43	129.35	131.14
4000	132.45	135.37	137.16
6000	135.97	138.89	140.68
8000	138.47	141.39	143.18
10000	140.41	143.33	145.12

In order to ensure stable radio communication and, as a result, high-quality data transmission, it is required that the signal level arriving at the receiver input exceed the noise level. We assume that the noise level for the 2–6 GHz ranges does not exceed minus 90 dBm and the optimal excess of the signal level over the noise level is 10 dB. Thus, the minimum signal power level at the receiver input is $P_r = -80$ dBm.

Let's give an example of calculating the transmitter power to organize a communication channel between a base station located on a tethered drone and a ground subscriber station. We assume that wireless communication is implemented at a frequency of 6000 MHz, the transmitter antenna with a circular radiation pattern has a gain $G_t = 6$ dBi, and the directional antenna of ground equipment has a gain G_r in the range of 15–25 dBi. Then from Table 2 and formula (3) we find that for reliable signal reception ($P_r = -80$ dBm) at a distance of 35 km, attenuation $\Pi = 138.89$ dB, the required transmitter power P_t is: 27.9–37.9 dBm.

3 Architecture of Wideband Wireless Network Based on Tethered Drone

The structure of a wideband wireless network based on a tethered drone consists of a radio cell, with a central base station (BS) equipped with an omnidirectional or sector antenna, onto which the antennas of subscriber stations (SS) are focused. In such a network, SS often do not have direct radio visibility with each other and are hidden from one another, with interaction occurring only through the BS. Topologically, the structure of such a network is star-shaped. International standards such as IEEE 802.11 [1] and its amendments have been developed for effective management of access to the wireless data transmission channel.

The next consideration is a wideband wireless network with a centralized management mechanism. The BS conducts a cyclic polling of SS according to a polling table, enabling more efficient use of network bandwidth, ensuring priority data transmission, and minimizing interference. The duration of one cycle is divided into two intervals. During the first interval (T_1), transmission occurs from the BS to the subscribers (downlink), and during the second interval (T_2), data is transmitted from the SS (uplink). The cycle duration, and thus the ratio of downlink/uplink transmission intervals, is set during the configuration stage, i.e., prior to the network's operation.

Each time interval (T_1, T_2) is divided into fixed duration time slots. Each slot contains a control interval (T_3) during which the polling table is transmitted, determining the sequence of data transmission to the SS. The sequence of data transmission from the subscriber station to the BS is also determined by this polling table. The number of slots allocated to each subscriber station may depend on their priority. Transition from one cycle to another is carried out through a guard interval (T_4), which is intended to complete the servicing of the most distant subscriber (calculated from the ratio of 3.5 microseconds per kilometer).

To numerically evaluate the performance characteristics of a wireless network based on a tethered drone (performed in the next section), we will assume the following:

– the cycle duration is 5000 ms;
– the slot duration is chosen as 250 ms;
– the duration of the control and guard intervals is 30 ms and 17.5 ms, respectively.

4   Performance Evaluation of Wireless Broadband Network Protocol Based on Tethered Altitude Unmanned Platform Using Batch Service Polling Model

In this section, we evaluate the performance of the wireless broadband network by analyzing the packet loss probability and the mean packet sojourn time in the system. As a mathematical model adequately describing the operation of broadband wireless networks with a centralized control mechanism, we consider the stochastic polling model with a single server and $N(N \geq 2)$ queues from Q_1 to Q_N (see Fig. 3).

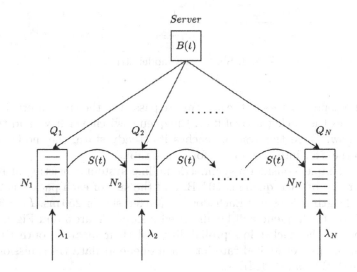

Fig. 3. Polling model

It is assumed that each queue receives a Poisson arrival flow of packets. The buffer size of each queue is limited by the value of N_i. In contrast to the known publications [4,13,14], batch packet servicing in the wireless network is considered. It is also assumed that servicing of packet batches is distributed according to an arbitrary distribution law $B(t)$, identical for all queues. The server switches between different queues for servicing at random times with a specified distribution function $S(t)$.

The objective is to determine the key performance characteristics of the wireless network, including the mean packet sojourn time and the packet loss probability in the system.

For complex models like the one presented in the current paper, obtaining analytical results using conventional mathematical queueing theory methods is difficult. Therefore, Monte Carlo simulation method is used to investigate the characteristics of the polling system with batch service. The simulation model was implemented in C++ using the OMNeT++ platform, which is a modular component library and a computer network simulation environment. The structure of the simulation model is illustrated in Fig. 4.

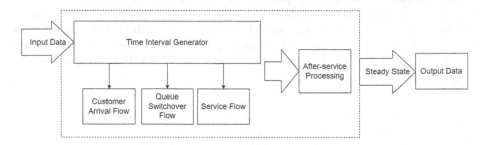

Fig. 4. Simulation model structure

The input parameters of the system are passed to the time interval generator which creates the packet arrival flow, the queue switchover flow, and the packet servicing flow. Once the system reaches its steady state, the model sends the results to the user interface in the form of a table.

As an example, consider a symmetric polling system consisting of 10 queues. The buffer size of each queue is 10 KB, and the size of each packet is 256 bytes. The cycle length is 5000 ms. Each slot has a fixed size of 250 ms. $T_3 = 30$ ms and $T_4 = 17.5$ ms. Each queue will be allocated 2 slots. Figure 5 and Fig. 6 show the dependence of the packet loss probability and their mean sojourn time in the system on the packet arrival rate into each queue at data transmission rates of 5 MB/s, 10 MB/s, and 20 MB/s.

It can be seen from Fig. 5 and Fig. 6 that as the data transmission rate increases, the number of serviced packets per queue visit by the server increases, reducing the packet loss probability and their mean sojourn time. Since the system is symmetric, the packet loss probability and mean sojourn time in different queues are the same.

Next, consider an example of an asymmetric polling system consisting of 15 queues. The other parameters of the system remain the same. Since the number of slots does not change, the first 5 queues will be allocated 2 slots, while the rest will only have 1 slot. Figure 7 and Fig. 8 also show the dependence of the packet loss probability and the packet mean sojourn time in the system on the packet arrival rate into each queue at three different data transmission rates. All these graphs have a similar shape to those in the case of symmetric systems.

However, since the queues in this example are allocated different numbers of slots, the packet loss probability and the mean packet sojourn time in different queues are different. An example is shown in Fig. 9 for the case of a data transmission rate of 5 MB/s. Since the first 5 queues are allocated 2 slots, the

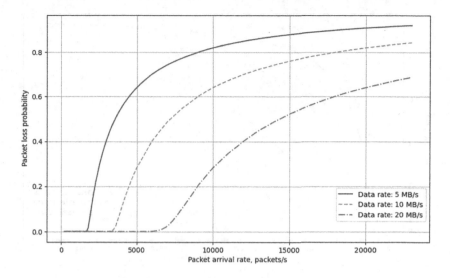

Fig. 5. Dependence of the packet loss probability on the packet arrival rate, number of queues – 10

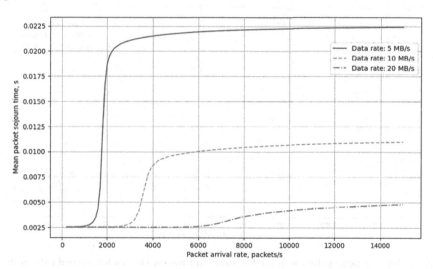

Fig. 6. Dependence of the mean packet sojourn time on the packet arrival rate, number of queues – 10

server will have more time to service packets in these queues, which will reduce the packet loss probability and the mean packet sojourn time in these queues compared to others.

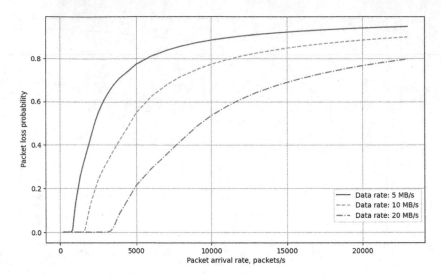

Fig. 7. Dependence of the packet loss probability on the packet arrival rate, number of queues – 15

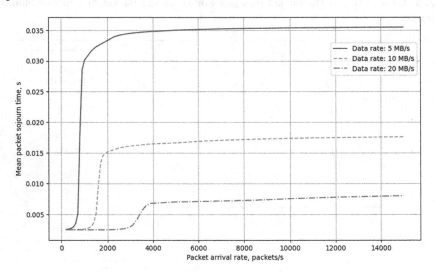

Fig. 8. Dependence of the mean packet sojourn time on the packet arrival rate, number of queues – 10

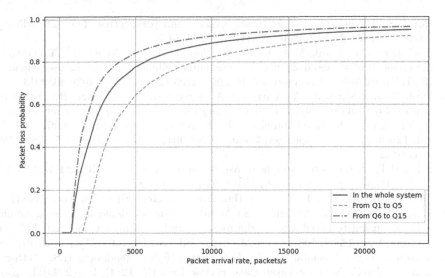

Fig. 9. Dependence of the mean packet loss probability in different queues on the packet arrival rate, number of queues – 15

5   Conclusion

This paper presents the results of performance evaluation of a broadband wireless network based on a tethered drone using an original stochastic polling model. In contrast to known works [4,13,14], a polling model with batch servicing of packets from wireless network subscribers is considered. A description of the protocol for interaction between the base station and subscriber devices of the broadband wireless network based on a tethered drone is given, allowing to obtain initial data for conducting numerical calculations of the simulation model of the polling system with batch servicing. The implemented simulation model allows to obtain the main performance characteristics of the broadband wireless network with a centralized control mechanism. Engineering calculations of the maximum radio coverage, required power, and antenna gain coefficients of the receiving and transmitting devices of the wireless network based on a tethered drone are also provided.

References

1. ANSI/IEEE Std 802.11, 1999 Edition: Wireless LAN Medium Access Control (MAC) and Physical Layer (PHY) Specifications (1999)
2. Arif, M., Kim, W.: Analysis of fluctuating antenna beamwidth in UAV-assisted cellular networks. Mathematics **11**(22), 4706 (2023). https://doi.org/10.3390/math11224706
3. Belmekki, B.E.Y., Alouini, M.S.: Unleashing the potential of networked tethered flying platforms: prospects, challenges, and applications. IEEE Open J. Veh. Technol. **3**, 278–320 (2022). https://doi.org/10.1109/OJVT.2022.3177946

4. Borst, S., Boxma, O.: Polling: past, present, and perspective. TOP **26**, 335–369 (2018). https://doi.org/10.1007/s11750-018-0484-5

5. Bushnaq, O.M., Kishk, M.A., Celik, A., Alouini, M.S., Al-Naffouri, T.Y.: Optimal deployment of tethered drones for maximum cellular coverage in user clusters. IEEE Trans. Wirel. Commun. **20**(3), 2092–2108 (2021). TWC.2020.3039013

6. Dinh, T.D., Vishnevsky, V., Larionov, A., Vybornova, A., Kirichek, R.: Structures and deployments of a flying network using tethered multicopters for emergencies. In: Vishnevskiy, V.M., Samouylov, K.E., Kozyrev, D.V. (eds.) DCCN 2020. LNCS, vol. 12563, pp. 28–38. Springer, Cham (2020). https://doi.org/10.1007/978-3-030-66471-8_3

7. Friis, H.T.: A note on a simple transmission formula. Proc. IRE **34**(5), 254–256 (1946). https://doi.org/10.1109/JRPROC.1946.234568

8. Johnson, R.: Antenna Engineering Handbook. McGraw-Hill, New York (1984)

9. Kishk, M., Bader, A., Alouini, M.S.: Aerial base station deployment in 6g cellular networks using tethered drones: the mobility and endurance tradeoff. IEEE Veh. Technol. Mag. **15**(4), 103–111 (2020). https://doi.org/10.1109/MVT.2020.3017885

10. Marques, M.N., Magalhães, S.A., Dos Santos, F.N., Mendonça, H.S.: Tethered unmanned aerial vehicles-a systematic review. Robotics **12**(4), 117 (2023). https://doi.org/10.3390/robotics12040117

11. Safwat, N.E.D., Hafez, I.M., Newagy, F.: 3d placement of a new tethered UAV to UAV relay system for coverage maximization. Electronics **11**(3), 385 (2022). https://doi.org/10.3390/electronics11030385

12. Vishnevsky, V., Selvamuthu, D., Rykov, V., Kozyrev, D., Ivanova, N., Krishnamoorthy, A.: Reliability Assessment of Tethered High-Altitude Unmanned Telecommunication Platforms: k-Out-of-n Reliability Models and Applications. Springer, Cham (2024). https://doi.org/10.1007/978-981-99-9445-8

13. Vishnevsky, V., Semenova, O.: Polling systems and their application to telecommunication networks. Mathematics **9**(2) (2021). https://doi.org/10.3390/math9020117

14. Vishnevsky, V., Vytovtov, K., Barabanova, E., Semenova, O.: Analysis of a map/m/1/n queue with periodic and non-periodic piecewise constant input rate. Mathematics **10**(10) (2022). https://doi.org/10.3390/math10101684

15. Vladimirov, S., Vishnevsky, V., Larionov, A., Kirichek, R.: The model of WBAN data acquisition network based on UFP. In: Vishnevskiy, V.M., Samouylov, K.E., Kozyrev, D.V. (eds.) DCCN 2020. LNCS, vol. 12563, pp. 220–231. Springer, Cham (2020). https://doi.org/10.1007/978-3-030-66471-8_18

16. Wang, Y., Zhang, B., Qin, S., Peng, J.: A channel rendezvous algorithm for multi-unmanned aerial vehicle networks based on average consensus. Sensors **23**(19), 8076 (2023). https://doi.org/10.3390/s23198076

17. Zhao, W., Zhang, J., Li, D.: Clustering and beamwidth optimization for UAV-assisted wireless communication. Sensors **23**(23) (2023). https://doi.org/10.3390/s23239614

18. Zhu, C., Shi, Y., Zhao, H., Chen, K., Zhang, T., Bao, C.: A fairness-enhanced federated learning scheduling mechanism for UAV-assisted emergency communication. Sensors **24**(5), 1599 (2024). https://doi.org/10.3390/s24051599

Author Index

V. M. Vishnevskiy et al. (Eds.): DCCN 2023, CCIS 2129, p. 113, 2024.
https://doi.org/10.1007/978-3-031-61835-2